NUMERICAL MODELING OF OCEAN CIRCULATION

Oceans are an essential part of the climate system. They dominate the hydrosphere and play a key role in natural climate variability. The modeling of ocean circulation is therefore very important, not only for its own sake, but also in terms of the prediction of weather patterns and the effects of climate change.

This book begins with an introduction to the basic computational techniques that are necessary for all models of the ocean and atmosphere, and the conditions they must satisfy. It contains descriptions of the workings of ocean models, the problems that must be solved in their construction, and how to evaluate computational results. Major emphasis is placed on those features that distinguish models of the ocean from other models in computational fluid dynamics, with the intention of examining ocean models critically, and determining what they do well and what they do poorly. Numerical analysis is introduced as needed, and exercises are included to illustrate major points. Additional resources are available at www.cambridge.org/9780521781824.

Developed from notes for a course taught in physical oceanography at the College of Oceanic and Atmospheric Sciences at Oregon State University, this book is ideal for graduate students of oceanography, geophysics, climatology and atmospheric science. It will also be of great interest to researchers in oceanography and atmospheric science.

ROBERT MILLER is a professor in the College of Oceanic and Atmospheric Sciences at Oregon State University, and is a member of the American Geophysical Union.

NUMERICAL MODELING OF OCEAN CIRCULATION

ROBERT N. MILLER
Oregon State University

CAMBRIDGE UNIVERSITY PRESS
Cambridge, New York, Melbourne, Madrid, Cape Town,
Singapore, São Paulo, Delhi, Mexico City

Cambridge University Press
The Edinburgh Building, Cambridge CB2 8RU, UK

Published in the United States of America by Cambridge University Press, New York

www.cambridge.org
Information on this title: www.cambridge.org/9780521781824

First published 2007

A catalogue record for this publication is available from the British Library

ISBN 978-0-521-78182-4 Hardback

Contents

Preface

This text grew out of notes for a course taught to graduate students in physical oceanography at the College of Oceanic and Atmospheric Sciences at Oregon State University. The students are typically in their second year of graduate school, having passed introductory courses in theoretical physical oceanography. This is the background assumed for the course. Most of the students at this point have seen some numerical analysis.

The course, and hence this text, is intended for all students of physical oceanography. Major emphasis is on those features that distinguish models of the ocean from other models in computational fluid mechanics. The intent is to examine ocean models critically, and determine what they do well and what they do poorly. We will ask when we can be confident that the model reflects nature, and when we can say that it is likely that we are looking at a feature of the model itself.

This is not a mathematics text as such, but it has a high mathematical content. Numerical analysis is introduced as needed. The reader may wish to consult supplementary references for basic numerical analysis of partial differential equations such as Sod (1985) (many typos, but reasonably current on fundamentals) or Richtmyer and Morton (1967), which is useful and commonly cited, though outdated. The reader might also find Isaacson and Keller (1966) or Allen *et al.* (1988) useful as general references. We will take examples from the two-volume work by Fletcher (1991) and the monograph by Leveque (1992). Durran (1999) contains useful and closely related material, from a slightly different viewpoint.

While implementation of detailed models has in the past been restricted to supercomputers at major computer laboratories, facilities at Oregon State University and many other institutions provide computing power of a magnitude available only at supercomputer centers a few years ago. Computational assignments and examples for this class were performed on computers at Oregon State University and at the National Center for Atmospheric Research (NCAR). Many, if not most, of the computing exercises presented here are now within the capabilities of common desktop computers and workstations. Examples and exercises are designed to illustrate questions of how models exhibit behavior typical of the real ocean and how and why they fail when they fail.

The scope of this text includes large-scale motions of the ocean in space and time. We consider the ocean as a shallow stratified fluid. By "shallow" we mean that we consider motions with horizontal scales much greater than the depth of the fluid. The motions which concern us evolve on timescales which are long compared to a day (or, more precisely, a "pendulum day," i.e., the period of a Foucault pendulum). Most of the motions considered here occur on large horizontal spatial scales. By "large" we will usually mean large compared to the distance the fastest internal wave travels in a day. While numerical ocean modeling would seem to be a diffuse subject that might not lend itself to a unified treatment in a single text, the majority of large-scale ocean models must deal with a number of common problems. Each way of solving these problems is a tradeoff, and a tactical choice that is suitable for one task may not be suitable for another.

Tides are not considered in this text, even though they are a prime example of shallow water motion. Tidal calculations for most purposes are performed without regard to stratification. More importantly, tidal calculations as performed in practice involve different computational strategies from those on which we focus here. Readers interested in the details of tidal calculations may refer to Foreman *et al.* (1993, 2000), Le Provost *et al.* (1994) or Ray (1993).

This work is intended as a text. It is not intended to review the state of the ocean modeling art. Rather its aim is to provide the student with the context in which discussion of numerical modeling is conducted. Attention is therefore confined to established facts and practices. For this reason, some very important topics, notably parameterization of turbulent processes, are omitted almost entirely, since even the basics of these topics are not well established in ocean modeling practice.

Chapters 2, 3 and 4 form the core of the ten-week course. There is usually time to treat one of the remaining chapters, and this is chosen according to class interests. Examples from the research literature are provided to illustrate the application of modeling techniques.

It is a pleasure to acknowledge the influence of many colleagues and mentors who have influenced my thinking on this topic. Conversations over the years with Andrew Bennett, Mark Cane, Michael Ghil, Ricardo Matano and Gary Sod, among others, have been particularly influential on my views of this subject. I would be especially pleased if mentors Alexandre Chorin and Allan Robinson were to see their fingerprints on this work. Linda Lamb's and Judy Scott's help with editing of the manuscript was invaluable, as was Dave Reinert's help with the figures. Acknowledgment is also made to the National Center for Atmospheric Research, which is sponsored by the National Science Foundation, for the computing time used in support of the classes and the compilation of results for this text.

1
Introduction

Numerical modeling has become an essential component of most research in physical oceanography. Once the domain of specialists, interpretation of model results has become part of the routine scientific practice. The highly schematic models of the 1970s and 1980s with their highly idealized geometry, coarse resolution and crudely parameterized small-scale processes could give only rough qualitative insights into applicability of theory. With the computing power available to nearly every scientist today, numerical models are expected to compare in detail with observation.

Field experiments are now designed with the intention of providing models with initial conditions, boundary conditions and verification data. Non-specialists, who, even five years ago, would not have given much thought to the implications of modeling results, now routinely ask themselves, "What do the models tell us?" Increasingly, this is the way oceanography is done.

Ocean modeling is closely related in method and spirit to atmospheric modeling, but atmospheric modeling was developed earlier, driven by the need for operational weather prediction. The basic methodology for weather forecasting was first set out by Richardson in a remarkable book (Richardson, 1965), first published in 1922. In that book, all of the steps for constructing a numerical weather forecast were set out in detail. A sample calculation was performed for two points in Europe, but the forecast turned out to be markedly different from the state of the atmosphere observed at the forecast time.

No automatic computing machinery was available to Richardson. He put considerable effort into the design of computing forms which would make the tabulation of data and intermediate results as convenient as possible. Actual calculations were performed with a slide rule and with log tables. In the preface to his book, dated October, 1921, he wrote:

> Perhaps some day in the dim future it will be possible to advance the computations faster than the weather advances and at a cost less than the saving to mankind due to the information gained. But that is a dream.

The next chapter follows in the spirit of Richardson's work. We examine schemes that appear reasonable but give entirely unreasonable results. We know now that the striking departure of Richardson's first forecast from reality was due to the failure of the calculation to satisfy a basic criterion for computational stability. Careful as Richardson was, he did not realize that his computing scheme, though evidently intuitively reasonable, was inherently unstable.

2

Some basic results from numerical analysis

2.1 Simple discretizations of a linear advection equation

We will use the simple one-dimensional advection equation with unit advection speed,

$$u_t + u_x = 0, \tag{2.1}$$

to study some of the fundamental consequences of discretization of partial differential equations. The exact solution to (2.1) can be found by examining curves in the $x - t$ plane of the form $t = x + x_0$. Along such curves,

$$\frac{\mathrm{d}u}{\mathrm{d}x} = u_t \frac{\mathrm{d}t}{\mathrm{d}x} + u_x = u_t + u_x.$$

So (2.1) states that along these curves u is constant. These curves are known as the *characteristics* of (2.1). We can use the characteristics to solve the initial value problem for (2.1). Consider the point $(\hat{x}, \hat{t})$ in the $x - t$ plane. The characteristic passing through $(\hat{x}, \hat{t})$ is

$$t = x + (\hat{t} - \hat{x}).$$

If we extrapolate this line back to the x-axis, we find, at $t = 0$, $x = -(\hat{t} - \hat{x})$. If we are given the initial condition $u(x, 0) = F(x)$, we have

$$u(\hat{x}, \hat{t}) = F(\hat{x} - \hat{t}) \quad \text{for any } \hat{x} \text{ and } \hat{t}.$$

One way of visualizing the foregoing is to note that, given the graph of the initial condition, the solution to (2.1) at time t can be constructed by translating the graph t units to the right.

For our first examples of different choices of discretization, we consider (2.1) with the periodic boundary condition $u(0, t) = u(2\pi, t)$. We choose

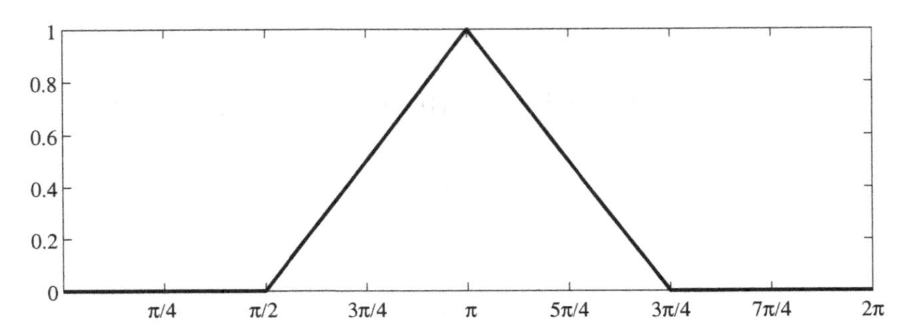

Fig. 2.1 Initial condition for advection equation (2.1).

a piecewise linear initial condition:

$$u(x,0) = \begin{cases} 0, & 0 \le x \le \pi/2, \\ (x-\pi/2)/(\pi/2), & \pi/2 < x \le \pi, \\ (3\pi/2 - x)/(\pi/2), & \pi < x \le 3\pi/2, \\ 0, & 3\pi/2 < x < 2\pi. \end{cases}$$

A graph of this function is shown in Figure 2.1 At time intervals of 2π, the solution should be identical to the initial condition. This is sometimes called the "one-dimensional color problem." One of the most obvious simple things to do is to approximate (2.1) in the interior by

$$\frac{u_j^{n+1} - u_j^n}{\Delta t} = -\left(\frac{u_{j+1}^n - u_{j-1}^n}{2\Delta x}\right), \tag{2.2}$$

where subscripts represent spatial grid position and superscripts represent time step. Periodic boundary conditions are applied to (2.2) at the boundary points. There are many ways to discretize (2.1), and others may seem more obvious, especially to the reader with a bit of experience in numerical solution of partial differential equations. Two things should be noted here about this particular choice. First, the spatial derivative of the solution at the point x_j is approximated by a difference between u_{j+1} and u_{j-1}. This is known as a centered difference scheme. Formal expansion of u in Taylor series about u_j easily shows the centered scheme to be more accurate than either of the so-called one-sided schemes, in which the derivative is approximated by $u_j - u_{j-1}$ or $u_{j+1} - u_j$. Second, (2.2) can be viewed as a method for finding an approximate solution to a system of coupled ordinary differential equations given by $\dot{u}_j = -(u_{j+1} - u_{j-1})/(2\Delta x)$. Methods of this form are sometimes placed in the general classification of "Method of Lines." The term comes from the fact that approximate solutions of the partial differential equation are computed on a family of lines in the $x - t$ plane. In this case, these lines are vertical, if we take x as the horizontal axis and t as the vertical axis.

The scheme (2.2) can be rearranged to form

$$u_j^{n+1} = u_j^n - \frac{\lambda}{2}(u_{j+1}^n - u_{j-1}^n), \tag{2.3}$$

where $\lambda = \Delta t/\Delta x$. If we choose 32 gridpoints and $\lambda = 0.5$, watch what happens at $t = 2\pi$; see Figure 2.2.

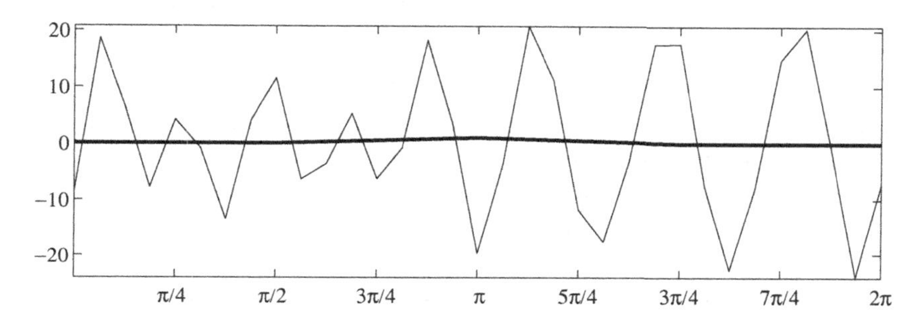

Fig. 2.2 Result of simple scheme for the advection equation (2.1). The heavy line depicts the true solution. The fine line depicts the computed approximate solution.

This result would certainly not lead us to put much faith in this scheme. The amplitude of the computed solution is an order of magnitude too great, and cursory inspection shows that it does not bear the slightest resemblance to the true solution. It turns out that fiddling with λ doesn't help.

Next try

$$\frac{u_j^{n+1} - u_j^n}{\Delta t} = -\left(\frac{u_j^n - u_{j-1}^n}{\Delta x}\right), \tag{2.4}$$

which can be written as

$$u_j^{n+1} = u_j^n - \lambda(u_j^n - u_{j-1}^n). \tag{2.5}$$

This is an example of a family of methods known as "upwind schemes," because the solution at a given point depends only on the initial condition at the point itself and points upwind relative to the advection velocity. The result for the same Δx and λ used in the above example is shown in Figure 2.3.

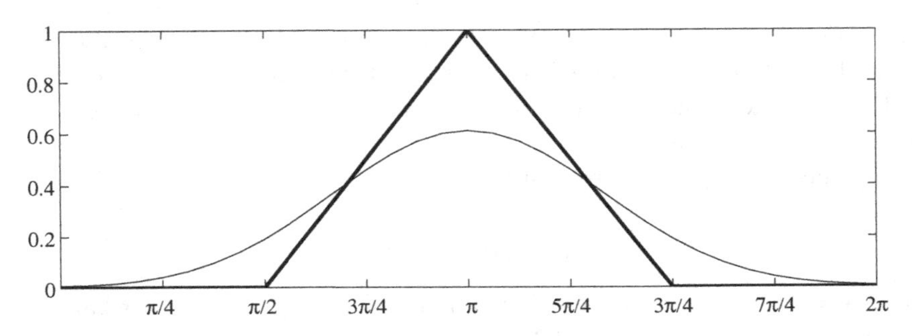

Fig. 2.3 Result of calculation with upwind scheme. Legend as in Fig. 2.2.

In this case, the amplitude is of the correct order of magnitude, and the approximate solution bears a clear resemblance to the true solution, but the spatial resolution is poor, i.e., the corners are rounded off. What happens if we increase the spatial resolution to 128 gridpoints, i.e., $\lambda = 2$? The result is shown in Figure 2.4. This is even worse than the result

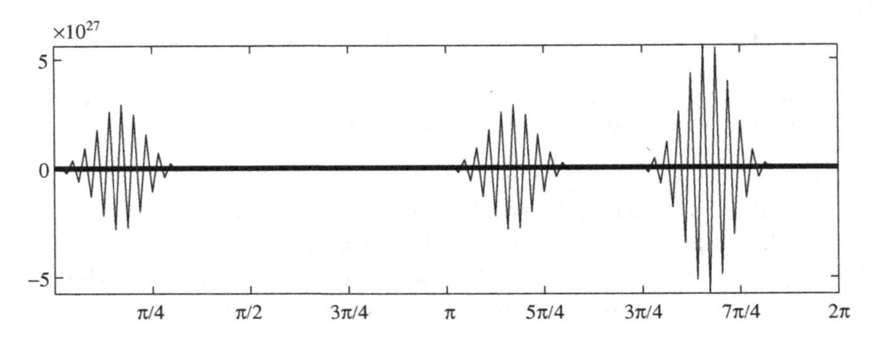

Fig. 2.4 Result of calculation with upwind scheme, $\lambda = 2$. Legend as before.

shown in Figure 2.2. The amplitude is 27 orders of magnitude too great, which is an unsatisfactory result by any standard. One might suspect that further continuation of this calculation could result in an overflow.

If we now return to the old value, i.e., $\lambda = 0.5$, by changing Δt appropriately, we see in Figure 2.5 that we have the better-resolved solution that we were seeking.

Next, let us investigate whether we can fix the centered scheme by trying

$$u_j^{n+1} = \frac{1}{2}(u_{j+1}^n + u_{j-1}^n) - \frac{\lambda}{2}(u_{j+1}^n - u_{j-1}^n). \tag{2.6}$$

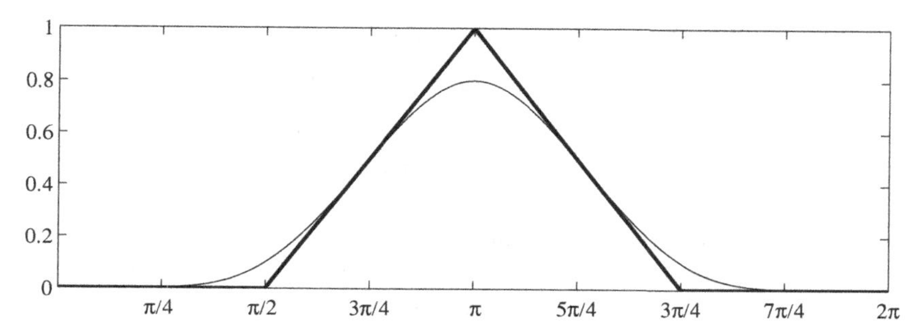

Fig. 2.5 Results of calculations with upwind scheme with $\Delta x = 2\pi/128$ and $\lambda = 0.5$. Legend as before.

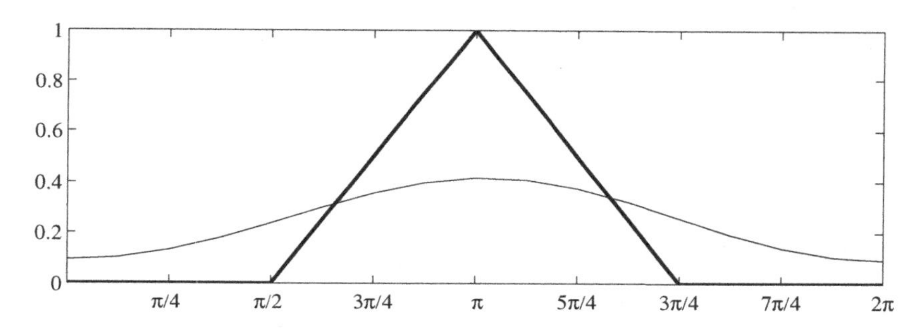

Fig. 2.6 Result of computation with the Lax–Friedrichs scheme with $\Delta x = 2\pi/32$ and $\lambda = 0.5$. Legend as before.

This is sometimes known as the Lax–Friedrichs scheme. The results of applying this scheme are shown in Figure 2.6. From that figure, we see that, as in the case of the upwind scheme, the result is of the correct order of magnitude, and has the correct general shape, but the approximate solution is unrealistically damped, to even a greater extent than the result from the upwind scheme with the same Δx and Δt. So the way we discretize something makes a great deal of difference. In the next section, we will use analytical techniques to investigate the effects of discretization.

2.2 Analysis of numerical results

2.2.1 Consistency, stability, convergence: the fundamentals

We began this chapter by taking a simple partial differential equation and approximating the partial derivatives by divided differences. All of the schemes we tried out were *consistent*, i.e., each individual difference quotient would converge to the derivative it was intended to approximate as Δt and Δx decreased. It is easy enough to formalize this. Let $v(x,t)$ be a solution of (2.1). We may write the scheme (2.3) as

$$\frac{u_j^{n+1}-u_j^n}{\Delta t}+\frac{u_{j+1}^n-u_{j-1}^n}{2\Delta x}=0.$$

We may now expand v in Taylor series:

$$\begin{aligned} v[j\Delta x,(n+1)\Delta t] &= v(j\Delta x,n\Delta t)+\Delta t v_t(j\Delta x,n\Delta t)+0(\Delta t^2)\\ v[(j+1)\Delta x,n\Delta t] &= v(j\Delta x,n\Delta t)+\Delta x v_x(j\Delta x,n\Delta t)\\ &\quad+\frac{1}{2}\Delta x^2 v_{xx}(j\Delta x+n\Delta t)+0(\Delta x^3), \end{aligned}$$

which gives us

$$\frac{v[j\Delta x,(n+1)\Delta t]-v(j\Delta x,n\Delta t)}{\Delta t}+\frac{v[(j+1)\Delta x,n\Delta t]-v[(j-1)\Delta x,n\Delta t]}{2\Delta x}=v_t+v_x+0(\Delta t)+0(\Delta x^2).$$

In the limit as Δt and Δx approach zero, the solution of the difference equations approaches the solution of the differential equation at any fixed x and t. We have seen that consistency is not enough. All of the schemes in Section 2.1 were consistent, yet some were obviously unsatisfactory.

The troubles illustrated in Figures 2.2 and 2.4 can be avoided if we choose schemes that are *stable*. A scheme for solution of an initial value problem is stable if small changes in the initial condition result in small changes in the result at some fixed time, say $t = 2\pi$, as in the cases shown in the previous section. To make this precise, we must say what we mean by "small."

We begin by writing the approximate solution after n time steps as the vector $\mathbf{v}^n$, whose jth component is the value v_j^n of the approximate solution at the jth grid point after n time steps. We decide whether a vector is large or small by assigning a *norm* to it. A norm of $\mathbf{v}$ is a scalar valued function written $\|\mathbf{v}\|$ which is positive definite, i.e.,

$$\|\mathbf{v}\| \geq 0, \tag{2.7}$$

with equality holding if and only if $\mathbf{v} = 0$ and obeys the triangle inequality

$$\|\mathbf{v} + \mathbf{w}\| \leq \|\mathbf{v}\| + \|\mathbf{w}\| \tag{2.8}$$

for any pair of vectors $\mathbf{v}$ and $\mathbf{w}$. A norm is a general notion of length. The ordinary Euclidean length of a vector, given by $(\sum_j v_j^2)^{1/2}$, is an example of a norm. There are times when other norms are more convenient to use or give rise to more enlightening results.

We may formulate a definition of stability in a number of ways. The following is standard for linear difference schemes.

Let $\mathbf{v}^n$ be the discrete solution at time $n\Delta t$, and let the difference scheme be given by

$$\mathbf{v}^{n+1} = L(\Delta t)\mathbf{v}^n.$$

All of the schemes in Section 2.1 can be written in this form. Our definition of stability will be given in terms of a norm of the matrix L. Here we assign a norm to a matrix by saying that the norm of a matrix L is large if there is some unit vector $\mathbf{v}$ such that the norm of the product $L\mathbf{v}$ is large, i.e., given a vector norm $\|\cdot\|$, the norm of the matrix L is the maximum over all vectors $\mathbf{u}$ of $\|L\mathbf{u}\|/\|\mathbf{u}\|$. This norm is referred to as the matrix norm *induced* by $\|\cdot\|$. The scheme is said to be stable if

$$\|L\| \leq 1 + 0(\Delta t).$$

Note that solutions to a stable scheme may grow at most exponentially. Fix some time $T = n\Delta t$, and let $L(\Delta t)$ be stable. Then

$$\mathbf{v}^n = L^n\mathbf{v}^0$$

and

$$\|\mathbf{v}^n\| = \|L^n\mathbf{v}^0\| \leq \|L\|^n\|\mathbf{v}^0\| \leq (1 + k\Delta t)^n\|\mathbf{v}_0\|$$

for some k, if Δt is sufficiently small. Therefore

$$\|\mathbf{v}^n\| = \left(1 + \frac{kT}{n}\right)^n \|\mathbf{v}_0\| \leq \mathrm{e}^{kT}\|\mathbf{v}_0\|.$$

On the other hand, if $\|L\| > 1$ independent of Δt, the solution will obviously blow up as $n \to \infty$.

For nonlinear systems, this definition is not so hard to generalize. We wish stability of a numerical scheme to mean the same thing as stability of anything else: the effect of small perturbations should remain bounded for finite time. So if $\mathbf{v}$ and $\mathbf{w}$ are solutions to a given difference scheme at time T with initial conditions $\mathbf{v}^0$ and $\mathbf{w}^0$ respectively, we wish $\|\mathbf{v} - \mathbf{w}\|$ to be bounded in terms of $\|\mathbf{v}^0 - \mathbf{w}^0\|$, independent of Δt. This condition is usually very hard to verify.

We have seen that the exact solution to (2.1) can be found by examining the characteristics, and we found that in this case, the solution to (2.1) at any point depends only on the initial value at a single point on the x-axis. We say the *domain of dependence* for the solution of an initial value problem at the point $(\hat{x}, \hat{t})$ is that set of points on the x-axis at which the initial values influence the solution at $(\hat{x}, \hat{t})$. In this simple case, the domain of dependence consists of a single point $x = \hat{x} - \hat{t}$. We may define a *numerical domain of dependence* at the point $(j\Delta x,\ n\Delta t)$ analogously. The Courant–Friedrichs–Lewy (CFL) theorem states that in order for a numerical method to be stable, the numerical domain of dependence must contain the domain of dependence for the differential equation. This condition is necessary but not sufficient.

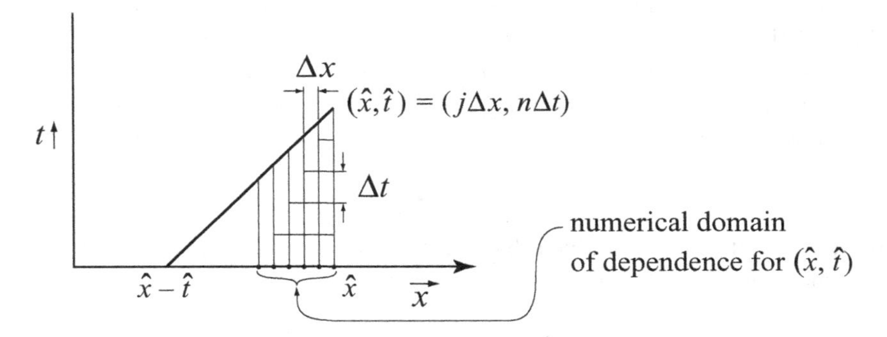

Fig. 2.7 Schematic diagram of the upwind scheme in the characteristic plane for $\lambda \approx 2$.

Let us consider the upwind scheme (2.4). The behavior of the scheme in the characteristic plane is illustrated in Figure 2.7. The leftmost point in the numerical domain of dependence is $j\Delta x - (n/\lambda)\Delta t$; the CFL criterion for stability is then $j\Delta x - (n/\lambda)\Delta t < j\Delta x - n\Delta t$, or $\lambda \leq 1$. So if $\lambda > 1$, the solution to the numerical scheme cannot converge to the solution of the PDE, no matter how small Δx and Δt are. This explains the behavior of the upwind scheme (2.4), but a similar analysis shows that the CFL criterion is satisfied for $\lambda < 1$ for scheme (2.2) also. A more delicate analysis is required to understand the behavior of (2.2).

Another way to investigate the stability of a linear scheme is by Fourier analysis. Consider the upwind scheme for the simple advection equation:

$$u_j^{n+1} = (1-\lambda)u_j^n + \lambda u_{j-1}^n, \tag{2.9}$$

where $\lambda = \Delta t/\Delta x$ as before. Now look for solutions of the form

$$u_j^n = \hat{U}^n e^{ikj\Delta x}. \tag{2.10}$$

We can derive a transformed scheme by substituting (2.10) into (2.9):

$$u_j^{n+1} = \left[(1-\lambda) + \lambda e^{-ik\Delta x}\right] \hat{U}^n e^{ikj\Delta x} \equiv \rho(k)\hat{U}^n e^{ikj\Delta x} = \rho(k)u_j^n,$$

so we may write

$$\hat{U}_{n+1} = \rho(k)\hat{U}_n.$$

If the transformed scheme turns out to be unstable, we can clearly construct an unstable solution of the original scheme. Stability of the transformed scheme is thus necessary for the stability of the original scheme. It is also sufficient in the case of a scalar equation, but not necessarily so for vector equations.

The expression $\rho(k)$ is called the *symbol* or the *amplification factor* of the difference scheme. We can check the stability of the scheme in wavenumber space by calculating

$$\begin{aligned}
\left|\rho(k)\right|^2 &= \left[(1-\lambda) + \lambda\cos k\Delta x\right]^2 + \lambda^2 \sin^2 k\Delta x \\
&= (1-\lambda)^2 + 2\lambda(1-\lambda)\cos k\Delta x + \lambda^2 \\
&= 1 - 2\lambda(1-\lambda)(1-\cos k\Delta x).
\end{aligned}$$

Now $0 \le 1 - \cos k\Delta x \le 2$, so the scheme is stable if $0 \le 2\lambda(1-\lambda) \le 1$. Examination of the graph of $\lambda(1-\lambda)$ (see Figure 2.8) shows that the scheme is stable for $0 \le \lambda \le 1$.

Consider now the centered scheme (2.3):

$$u_j^{n+1} = u_j^n - \frac{\lambda}{2}(u_{j+1}^n - u_{j-1}^n).$$

Putting $u_j^n = \hat{U}^n e^{ijk\Delta x}$ yields

$$u_j^{n+1} = [1 - i\lambda \sin k\Delta x]u_j^n,$$

so $\left|\rho\right|^2 = 1 + \lambda^2 \sin^2 k\Delta x$ and the scheme is unstable for any λ. Intuitively, one might argue that the scheme is stable, since $\sin k\Delta x \approx k\Delta x$

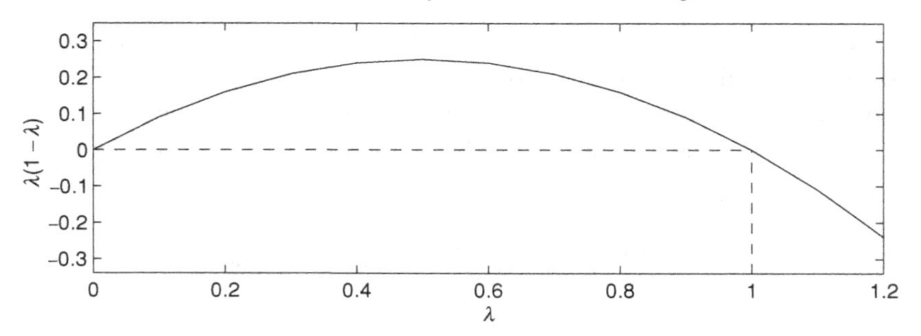

Fig. 2.8 Graph of $\lambda(1-\lambda)$.

so $\lambda^2 \sin^2 k\Delta x = 0(k^2\Delta t^2) = 0(k\Delta t)$, but this is only true for $k\Delta x << 1$, and there will always be computational modes with high wavenumber, introduced by roundoff if nothing else. So the simple centered scheme is unstable, even when the CFL condition is met (see Exercise 2.2).

Consider now the leapfrog scheme:

$$\frac{u_j^{n+1} - u_j^{n-1}}{2\Delta t} + \frac{u_{j+1}^n - u_{j-1}^n}{2\Delta x} = 0,$$

so

$$u_j^{n+1} = u_j^{n-1} - \lambda(u_{j+1}^n - u_{j-1}^n). \tag{2.11}$$

For purposes of analysis, we may write this as a vector system by introducing a new variable $v_j^n = u_j^{n-1}$, so

$$\begin{aligned} u_j^{n+1} &= v_j^n - \lambda(u_{j+1}^n - u_{j-1}^n) \\ v_j^{n+1} &= u_j^n. \end{aligned}$$

Writing $u_j^n = \hat{u}^n(k)\mathrm{e}^{\mathrm{i}jk\Delta x}$ and $v_j^n = \hat{v}^n(k)\mathrm{e}^{\mathrm{i}jk\Delta x}$ gives

$$\begin{pmatrix} \hat{u}^{n+1} \\ \hat{v}^{n+1} \end{pmatrix} = \begin{pmatrix} -2i\lambda \sin k\Delta x & 1 \\ 1 & 0 \end{pmatrix} \begin{pmatrix} \hat{u}^n \\ \hat{v}^n \end{pmatrix} \equiv \hat{G} \begin{pmatrix} \hat{u}^n \\ \hat{v}^n \end{pmatrix}. \tag{2.12}$$

A necessary but not sufficient condition that the scheme be stable is that no eigenvalue of $\hat{G}$ has absolute value $> 1 + 0(\Delta t)$; the quantity $\max_i |\mu_i(\hat{G})|$, where $\mu_i(\hat{G})$ is the ith eigenvalue, is known as $\sigma(\hat{G})$, the *spectral radius.* The requirement that $\sigma(\hat{G}) \le 1 + 0(\Delta t)$ is known as the *von Neumann condition.* In this case the eigenvalues of $\hat{G}$ satisfy

$$\mu(\mu + 2\mathrm{i}\lambda \sin k\Delta x) - 1 = 0,$$

so

$$\mu = -\mathrm{i}\lambda \sin k\Delta x \pm \sqrt{1 - \lambda^2 \sin^2 k\Delta x.}$$

For $\lambda < 1, |\mu|^2 < 1$, the von Neumann condition is satisfied and the method is stable. The problem is a bit more delicate when $\lambda = 1$. If $\lambda = 1$ and $k\Delta x = -\pi/2$,

$$\hat{G} = \begin{pmatrix} 2\mathrm{i} & 1 \\ 1 & 0 \end{pmatrix} = \begin{pmatrix} \mathrm{i} & 1 \\ 1 & 0 \end{pmatrix} \begin{pmatrix} \mathrm{i} & 1 \\ 0 & \mathrm{i} \end{pmatrix} \begin{pmatrix} 0 & 1 \\ 1 & -\mathrm{i} \end{pmatrix}.$$

Note that

$$\begin{pmatrix} \mathrm{i} & 1 \\ 1 & 0 \end{pmatrix} \begin{pmatrix} 0 & 1 \\ 1 & -\mathrm{i} \end{pmatrix} = I,$$

where $\begin{pmatrix} \mathrm{i} & 1 \\ 0 & \mathrm{i} \end{pmatrix}$ is the *Jordan Canonical Form* of $\hat{G}$, and

$$\hat{G}^n = \begin{pmatrix} \mathrm{i} & 1 \\ 1 & 0 \end{pmatrix} \begin{pmatrix} \mathrm{i} & 1 \\ 0 & \mathrm{i} \end{pmatrix}^n \begin{pmatrix} 0 & 1 \\ 1 & -\mathrm{i} \end{pmatrix} = \begin{pmatrix} \mathrm{i} & 1 \\ 1 & 0 \end{pmatrix} \mathrm{i}^n \begin{pmatrix} 1 & -\mathrm{i} \\ 0 & 1 \end{pmatrix}^n \begin{pmatrix} 0 & 1 \\ 1 & -\mathrm{i} \end{pmatrix}.$$

Now

$$\begin{pmatrix} 1 & -\mathrm{i} \\ 0 & 1 \end{pmatrix} = I - \begin{pmatrix} 0 & \mathrm{i} \\ 0 & 0 \end{pmatrix}, \text{ and } \begin{pmatrix} 0 & \mathrm{i} \\ 0 & 0 \end{pmatrix}^2 = 0,$$

so the binomial expansion

$$\left[I - \begin{pmatrix} 0 & \mathrm{i} \\ 0 & 0 \end{pmatrix} \right]^n = I - n\mathrm{i} \begin{pmatrix} 0 & 1 \\ 0 & 0 \end{pmatrix} = \begin{pmatrix} 1 & -n\mathrm{i} \\ 0 & 1 \end{pmatrix},$$

and therefore

$$\hat{G}^n = \mathrm{i}^n \begin{pmatrix} 1+n & -in \\ -in & 1-n \end{pmatrix},$$

which cannot be bounded independent of n, so the leapfrog scheme with $\lambda = 1$ is unstable. The leapfrog scheme with $\lambda = 1$ satisfies the CFL condition and the von Neumann condition, but is still unstable; see Exercise 2.5.

What we really want from a scheme is *convergence*. Assume $v(x, T)$ is the true solution to the given partial differential equation at time T with initial condition $v(x, 0)$; note that this presupposes that such a solution exists, is unique, and is sufficiently smooth. Put $x = j\Delta x, T = n\Delta t$. Now let $\mathbf{u}^n$, the vector with components u_j^n, be the solution to a difference scheme of the form

$$\mathbf{u}^{k+1} = L\mathbf{u}^k.$$

Then the scheme is said to be *convergent of order (p,q)* in the norm $\|\cdot\|$ if

$$\|v(x,T) - \mathbf{u}^n\| = O(\Delta x^p) + O(\Delta t^q) \text{ as } \Delta t,\ \Delta x \to 0. \tag{2.13}$$

Note that in the above limit $n \to \infty$ as $\Delta t \to 0$. Note also that the left-hand side of (2.13) contains an abuse of notation since v is a function defined on an interval in space while $\mathbf{u}^n$ is defined on a discrete set of points. The expression $v(x,T) - \mathbf{u}^n$ therefore presupposes sampling v or interpolating $\mathbf{u}^n$. The relation between consistency, stability and convergence is given by the *Lax Equivalence Theorem*:

For linear systems, a difference scheme is convergent if and only if it is consistent and stable.

Proof of this theorem can be found in Sod (1985) or Richtmyer and Morton (1967).

2.3 Implicit methods

Schemes of the form $L\mathbf{u}^{n+1} = M\mathbf{u}^n$ for nontrivial L are known as implicit schemes. The simplest implicit scheme for (2.1) can be constructed by calculating the spatial difference quotients at the advanced time:

$$\frac{u_j^{n+1} - u_j^n}{\Delta t} = -\left[\frac{u_{j+1}^{n+1} - u_{j-1}^{n+1}}{2\Delta x}\right]$$

$$\frac{1}{2}\lambda(u_{j+1}^{n+1} - u_{j-1}^{n+1}) + u_j^{n+1} = u_j^n.$$

In this simple system, we can solve for ρ:

$$(\lambda \mathrm{i} \sin k\Delta x + 1)\rho = 1,$$

so

$$\rho = (1 + \lambda \mathrm{i} \sin k\Delta x)^{-1},$$

and the scheme is thus *unconditionally stable*, since the von Neumann condition is sufficient for scalar equations. Why do you suppose we don't use implicit methods much?

In nonlinear equations, where we have $u_t = f(u, u_x, u_{xx})$, these methods are not so convenient. Consider, e.g., Burgers' equation:

$$u_t + \left(\frac{1}{2}u^2\right)_x = \frac{1}{\mathrm{Re}}u_{xx},$$

Table 2.1 Table of difference schemes for the simple advection equation $u_t + u_x = 0$

Scheme	Definition	Stability
Simple centered	$u_j^{n+1} = u_j^n - \frac{\lambda}{2}(u_{j+1}^n - u_{j-1}^n)$	unstable
Upwind	$u_j^{n+1} = u_j^n - \lambda(u_j^n - u_{j-1}^n)$	$\lambda < 1$
Lax–Friedrichs	$u_j^{n+1} = \frac{1}{2}(u_{j+1}^n + u_{j-1}^n) - \frac{\lambda}{2}(u_{j+1}^n - u_{j-1}^n)$	$\lambda \leq 1$
Leapfrog	$u_j^{n+1} = u_j^{n-1} - \lambda(u_{j+1}^n - u_{j-1}^n)$	$\lambda < 1$
Implicit	$\frac{1}{2}\lambda(u_{j+1}^{n+1} - u_{j-1}^{n+1}) + u_j^{n+1} = u_j^n$	unconditional

where Re is the Reynolds number. A simple first-order implicit method might be written

$$u_j^{n+1} = u_j^n - \frac{\lambda}{4}[(u_{j+1}^{n+1})^2 - (u_{j-1}^{n+1})^2] + \frac{\lambda}{\text{Re}} \cdot \frac{1}{\Delta x}[u_{j-1}^{n+1} - 2u_j^{n+1} + u_{j+1}^{n+1}]$$

and we would have a set of nonlinear equations to solve each time step. The problem becomes much worse for applications to two and three space dimensions.

A summary of the simple difference schemes discussed up to this point appears in Table 2.1.

2.4 Dissipation and dispersion

The Lax–Friedrichs scheme (2.6),

$$u_j^{n+1} = \frac{1}{2}(u_{j+1}^n + u_{j-1}^n) - \frac{\lambda}{2}(u_{j+1}^n - u_{j-1}^n),$$

was observed earlier to be stable for $\lambda = 0.5$. One way to look at this is to rewrite it as

$$u_j^{n+1} = u_j^n + \frac{\Delta x^2}{2}\left(\frac{u_{j+1}^n - 2u_j^n + u_{j-1}^n}{\Delta x^2}\right) - \frac{\lambda}{2}(u_{j+1}^n - u_{j-1}^n).$$

Note that

$$\frac{u_{j+1}^n - 2u_j^n + u_{j-1}^n}{\Delta x^2} = \frac{\partial^2 u}{\partial x^2}(j\Delta x, n\Delta t) + 0(\Delta x^2),$$

so the Lax–Friedrichs scheme is actually an approximation to the equation

$$u_t + u_x = \left(\frac{\Delta x}{2\lambda}\right) u_{xx},$$

i.e., a diffusion term of $O(\Delta x)$ has been added.

Now consider the leapfrog scheme, for the stable case $\lambda < 1$:

$$u_j^{n+1} = v_j^n - \lambda(u_{j+1}^n - u_{j-1}^n),$$
$$v_j^{n+1} = u_j^n.$$

Its Fourier transform is given by

$$\begin{pmatrix} \hat{u}^{n+1} \\ \hat{v}^{n+1} \end{pmatrix} = \begin{pmatrix} -2\mathrm{i}\lambda \sin k\Delta x & 1 \\ 1 & 0 \end{pmatrix} \begin{pmatrix} \hat{u}^n \\ \hat{v}^n \end{pmatrix}.$$

The eigenvalues are

$$\mu_\pm = -\mathrm{i}\lambda \sin k\Delta x \pm \sqrt{1 - \lambda^2 \sin^2 k\Delta x}.$$

The eigenvectors are

$$\begin{pmatrix} \mu_\pm \\ 1 \end{pmatrix}. \tag{2.14}$$

Therefore, there are two wavelike solutions with wavenumber k, each of which has the form

$$u_j^n \propto \mu_\pm^n \, \mathrm{e}^{\mathrm{i}jk\Delta x} \begin{pmatrix} \mu_\pm \\ 1 \end{pmatrix}, \tag{2.15}$$

where $\propto$ denotes proportionality. Now

$$|\mu_\pm| = 1, \text{so } \mu_\pm = \mathrm{e}^{\mathrm{i}\Theta_\pm},$$

where $\Theta_\pm$ is given by

$$\begin{aligned} \tan \Theta_\pm &= \frac{\lambda \sin k\Delta x}{\mp\sqrt{1 - \lambda^2 \sin^2 k\Delta x}} = \frac{\sin k\Delta x}{k\Delta x} \left(\frac{k\Delta t}{\mp\sqrt{1 - \lambda^2 \sin^2 k\Delta x}} \right) \\ &\equiv \tan(-\omega \Delta t). \end{aligned} \tag{2.16}$$

Such an equation is known as a dispersion relation. The exact dispersion relation is $\Theta = -k\Delta t$, i.e., $\omega = k$. We will investigate this in a moment. First let us settle the matter of which root we want.

From the dispersion relation, we see that there are two solutions, and only one is physically reasonable: that is the Θ_+ root, since the wave corresponding to the Θ_- root goes the wrong way; note the $\mp$ in front of the radical in the dispersion relation, so choosing the negative root in the denominator of the right-hand side of (2.16) corresponds to Θ_+, the right-going wave. How shall we discriminate?

The answer is: by the initial conditions. Note that we need initial conditions for both u and v, and v was constructed so that $v_j^n = u_j^{n-1}$. This corresponds to the fact that the leapfrog scheme needs to be initialized at two time levels; the initial values alone are not sufficient to specify the solution to the entire problem. Presumably, given initial conditions at time $t = 0$, we will do something reasonable at $t = \Delta t$ in order to start the leapfrog calculation. So for fixed λ and a given wavenumber k, if the initial component at that wavenumber has unit amplitude, we will choose initial conditions for that component to be

$$v_j^0 = \mathrm{e}^{\mathrm{i}jk\Delta x}; \quad u_j^0 = \mathrm{e}^{\mathrm{i}\Theta_+}\mathrm{e}^{\mathrm{i}jk\Delta x}.$$

With the initial conditions so constructed, the solution to the difference equation will be

$$u_j^n = \mathrm{e}^{\mathrm{i}(jk\Delta x + n\Theta_+)}.$$

Now what about the dispersion relation? A wave is said to be *dispersive* if the phase speed depends nontrivially upon the wavenumber (i.e., waves with different wavelengths travel at different speeds). The underlying equation $u_t + u_x = 0$ does not give rise to dispersive waves – all waves travel at unit speed, independent of wavelength. The computed solutions, however, are dispersive. The above dispersion relation implies that

$$C_{\text{phase}} = \frac{\omega}{k} \approx \frac{\sin k\Delta x}{k\Delta x} \cdot \frac{1}{\sqrt{1 - \lambda^2 \sin^2 k\Delta x}} \quad \text{for small } k\Delta x. \tag{2.17}$$

For small $k\Delta x$ (i.e., wave well resolved), $\omega/k \approx 1$. When $k\Delta x = \pi$, $\sin k\Delta x = 0$ and $\tan\Theta = 0$, so $\mu_\pm = \pm 1$. For $k\Delta x > \pi$, the computational waves alias into smaller k, and can appear to be going in the wrong direction. For a single component, the result is phase error: the solution will have the correct amplitude but the wrong phase at some fixed time $T > 0$. For fixed k, letting $\Delta x \to 0$ (and thus letting $\Delta t \to 0$ in order to keep λ constant) results in the phase error decreasing as $\sin(k\Delta x)/k\Delta x \to 1$. Of course, as Δx decreases, more wavenumbers will be resolved, and the phase error will always be large at the largest wavenumbers. So the leapfrog scheme is not dissipative, but it does introduce dispersion into a nondispersive problem. The physically meaningful dispersion curve, i.e., the "+" solution to (2.16), is shown in Figure 2.9, along with the exact dispersion relation $\omega\Delta t = k\lambda\Delta x$. This figure illustrates the content of (2.17), i.e., the speed of well-resolved waves will be reasonably well simulated by the leapfrog scheme. The dispersion relation for the artificial computational mode would be

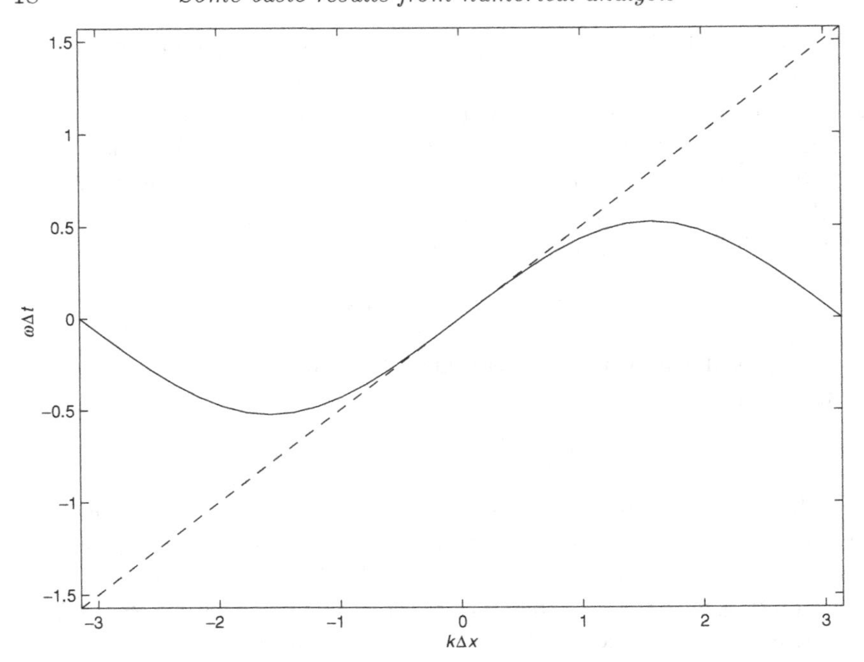

Fig. 2.9 Dispersion relation for the leapfrog scheme for the simple advection equation. The solid curve shows the physically meaningful wave mode. The dispersion relation for the computational mode is given by the negative of the solid curve. The dashed line depicts the exact dispersion curve for the partial differential equation. In this example $\lambda = \Delta t/\Delta x = 0.5$.

represented according to (2.16) as the negative of the solid curve in Figure 2.9.

There is another practical problem with the leapfrog scheme. It can be seen from the form of the scheme that the solution u_j^{n+1} does not depend on u_j^n, u_{j+1}^{n-1}, u_{j-1}^{n-1} and so forth. This is illustrated schematically in Figure 2.10, which illustrates the fact that at any given time level, solutions with odd and even space indices do not depend on one another. The even and odd solutions may diverge due to high wavenumber perturbations, and thus give rise to an artificially rough field. This phenomenon is sometimes known as "checkerboarding." For this reason, it is common practice to stop long leapfrog integrations at some time interval (usually empirically determined) and combine the two solutions. Substituting a single step of another, perhaps less accurate, method at some time interval is another common strategy used to mitigate the effects of checkerboarding.

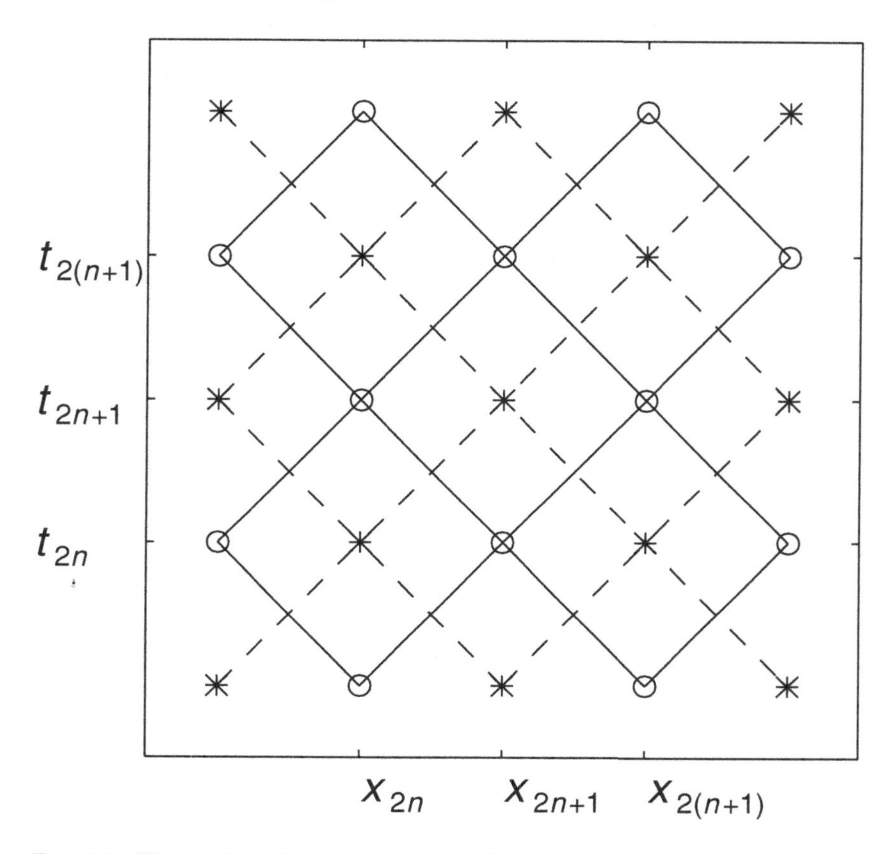

Fig. 2.10 Illustration of domains of dependence of the leapfrog scheme in the characteristic plane. The solution at points denoted by "∗" is independent of values of the solution at points denoted by "○".

The effect of checkerboarding is reflected in the two solutions noted in (2.15) and depicted in Figures 2.11 and 2.12. Panel (a) of each figure shows the physically reasonable solution. In Figure 2.11(a) the solution at time level 1, shown as the solid curve, is similar to the solution at time level zero, shown as the dashed curve, but displaced slightly to the right, reflecting rightward propagation of the wave at successive time steps. Panel (b) of each figure shows the computational mode. Initially (see Figure 2.11(b)), the solutions at two successive time steps are nearly negatives of one another.

The computed solutions for the two modes after one-quarter period are shown in Figure 2.12. In both cases, the initial condition is depicted as a curve in alternating dots and dashes. The "+" solution, shown on the left, behaves as expected. Both the final solution and the solution lagged by one time step propagate unchanged in shape to the right. In the computational mode, the pattern propagated to the left. We would not deliberately set up the computational mode as an initial condition, but if it is introduced by noise, it will persist.

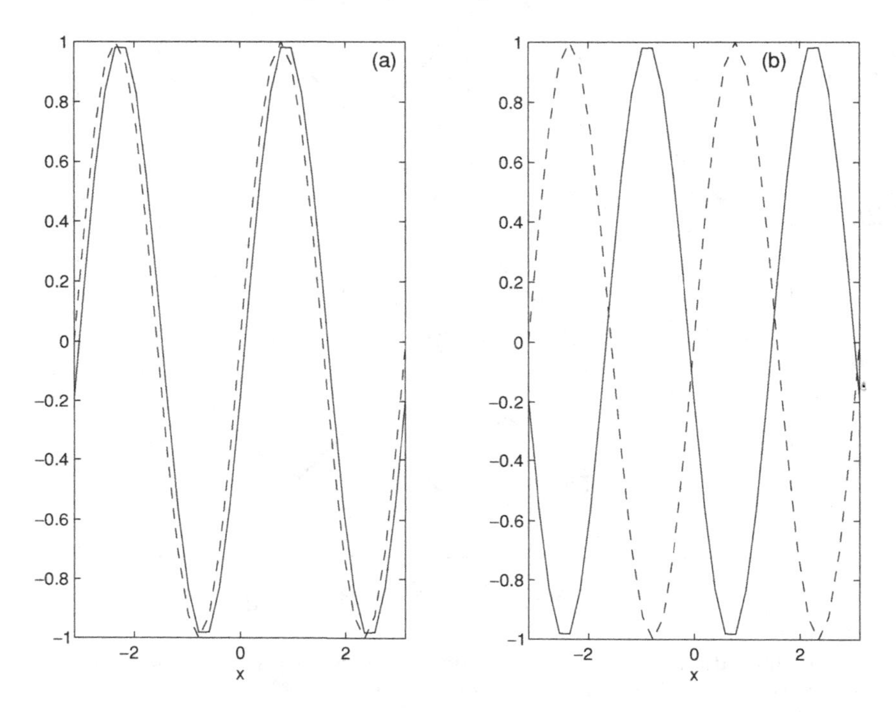

Fig. 2.11 Initial conditions for the two wave modes of the leapfrog scheme for the simple advection equation with periodic boundary conditions. $\lambda = 0.5$. The initial condition, shown as dashed line, is given by $\sin 2x$. The first time level is shown as solid line. The two time levels are defined as in (2.14). (a) "+" root, the physical mode; (b) "−" root, the computational mode.

2.5 A (very) brief introduction to finite-element methods

All of the methods we have seen so far have been derived by using finite differences to approximate derivatives. This seems an obvious approach,

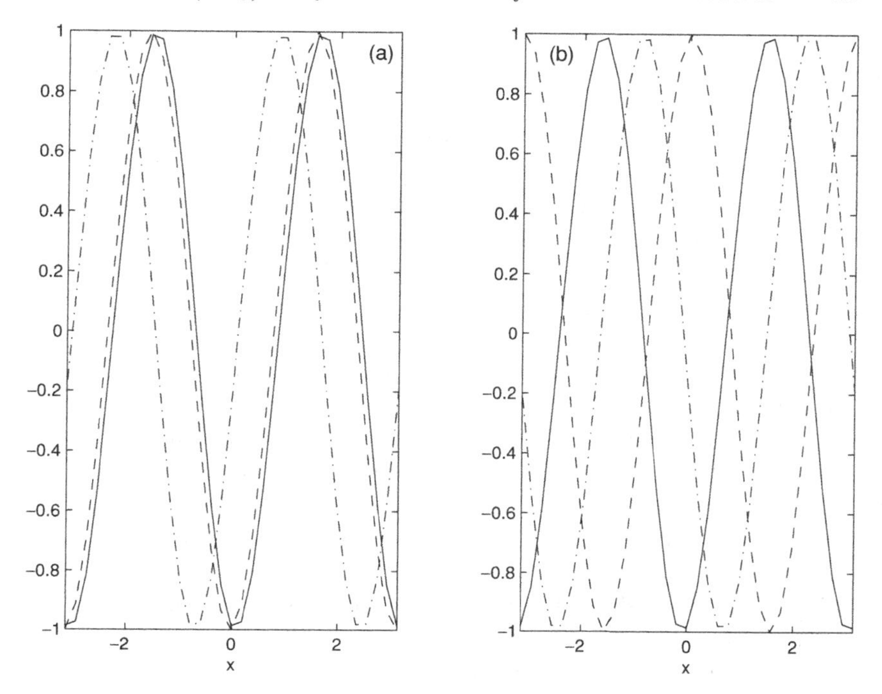

Fig. 2.12 Solution after 1/4 period for the two wave modes of the leapfrog scheme for the simple advection equation with periodic boundary conditions. $\lambda = 0.5$ and initial conditions as shown in Figure 2.11. Solid curves: solutions at 1/4 period. Dashed curves: lagged solutions, one step before the time of the solid curve. Dash-dotted curves: initial solutions at time step 1. (a) "+" root, the physical mode; (b) "−" root, the computational mode.

but it is not the only one. We might also calculate approximate solutions by dividing the domain up into pieces, and approximating the solution on each piece as a function of some specified form, say, a polynomial. This amounts to requiring our approximate solution to be a member of a finite-dimensional subspace of the infinite space of possible solutions. We then choose our approximate solution to be the function which lies in that predetermined finite-dimensional vector space and is as close as possible, in some reasonable sense, to the true solution.

One way to begin deriving a workable method using this technique is to choose a decomposition of the domain into pieces, which are called "elements," along with basis functions, sometimes called "shape functions" for the finite-dimensional approximation space. Suppose the basis consists of functions $\phi_i, i = 1, \ldots, N$. Note first that all of the problems we have encountered so far (e.g., (2.1)) can be written formally as rootfinding problems, i.e., find a function u such that $\mathbf{F}(u) = 0$. We seek an approximate solution $u^* = \sum_{i=0}^{N} \alpha_i \phi_i$ that is as close as possible to the true solution u. Substitution of the approximate solution u^* into the original equation will result in $\mathbf{F}(u^*) = r$, for some nonzero function r called the *residual.* In general, for a given finite basis set ϕ_i we cannot choose coefficients so that $r = 0$, so we choose a set of weighting functions $w_i, i = 1, \ldots, N$ so that $\int r w_i \, \mathrm{d}x = 0$ for $i = 1, \ldots, N$. General methods of this form are known as *weighted residual* methods, since the integral $\int r w_i \, \mathrm{d}x$ can be seen as a weighted average of the residual r, with the functions w_i, also known as *test functions*, acting as weights. A simple choice is $w_i = \phi_i$, resulting in $\int r \phi_j \, \mathrm{d}x \equiv (\phi_j, \mathbf{F}(u^*)) = 0$ for $j = 1, \ldots, N$. The expression $(\cdot, \cdot)$ denotes the inner product, analogous to the dot product of finite-dimensional vectors. This is equivalent to specifying the approximate solution to be the orthogonal projection of the true solution on the approximation space. Methods of this type are known as *Galerkin methods.*

As an example, consider the simple advection equation in one space dimension (2.1). For the moment consider an unbounded domain in space. Our approximate solution is given by

$$u^* = \sum_{i=0}^{N} u_i(t) \phi_i(x), \tag{2.18}$$

where the u_i are functions of time only. The basis functions ϕ_i will be chosen presently. Next, we divide the domain into equal intervals of length h, and choose piecewise linear basis functions ϕ_i:

$$\phi_i, i = 1, \ldots = \begin{cases} \frac{1}{h}[x - (i-1)h], & (i-1)h \leq x \leq ih, \\ \frac{1}{h}[(i+1)h - x], & ih \leq x \leq (i+1)h, \\ 0, & \text{otherwise}. \end{cases} \tag{2.19}$$

This is the space of all piecewise linear functions with corners at $x_i = ih$. By the foregoing, the Galerkin approximation to (2.1) is

$$\sum_j \dot{u}_j \int \phi_i \phi_j \mathrm{d}x + u_j \int \phi_i \phi_{jx} \mathrm{d}x \equiv \sum_j \dot{u}_j (\phi_i, \phi_j) + u_j (\phi_i, \phi_{jx}) = 0 \tag{2.20}$$

for all i. Note that the integral (ϕ_i, ϕ_{jx}) can be evaluated despite the fact that the basis functions are not differentiable even once.

The resulting semidiscrete approximate equations are

$$\frac{h}{6}\dot{u}_{i-1} + \frac{2h}{3}\dot{u}_i + \frac{h}{6}\dot{u}_{i+1} = -\left(\frac{1}{2}\right)(u_{i+1} - u_{i-1}). \tag{2.21}$$

Unlike most of the finite-difference methods we have examined so far, (2.21) is implicit. This is not serious in this case, since the system (2.21) is tridiagonal and easy to solve. Tridiagonal systems of order n can be solved in order n steps using the well-known Thomas algorithm (see, e.g., Isaacson and Keller, 1966; Allen *et al.*, 1988). Unfortunately, the Thomas algorithm does not parallelize naturally, but there are ways around that. Since the 1970s, tridiagonal systems have been considered essentially explicit for most intents and purposes. We know that the continuous system admits wave solutions. The semidiscrete scheme also has wavelike solutions of the form

$$u_j = \exp[i(jkh - \omega t)].$$

These waves satisfy the dispersion relation

$$\omega = \frac{3}{2 + \cos(kh)} \frac{\sin(kh)}{h}, \tag{2.22}$$

which approximates the true dispersion relation to fourth order in h. This reflects the general experience that finite-element methods are very efficient for advection problems. Difference methods of similar complexity are usually only second-order accurate.

We could also choose the test functions $w_j = \delta(x - x_j)$, $j = 1, \ldots$. This amounts to demanding that the approximate solution be an exact solution to (2.1) at $x = x_j$, $j = 1, \ldots$. Methods thus derived are known as *collocation methods.* We can devise an example of a collocation method for (2.1) by choosing the elements to be intervals of length h, and piecewise linear basis functions ϕ_i defined as before by (2.19). If the test functions are given by

$$w_j = \delta\left[x - \left(j + \frac{1}{2}\right)h\right], \tag{2.23}$$

we arrive at the scheme

$$\frac{\dot{u}_{j+1} + \dot{u}_j}{2} + \frac{u_{j+1} - u_j}{h} = 0, \tag{2.24}$$

which leads to the approximate dispersion relation

$$\omega = \frac{\tan(kh/2)}{kh/2}. \tag{2.25}$$

This approximation is only second-order accurate.

In the first example, we constructed the basis of our approximation space by dividing the domain into intervals and then choosing basis functions that were supported on each interval. A more accurate scheme might be derived by choosing smaller intervals and generating the basis functions in the same way. We could also keep the same intervals and choose basis functions that might yield better local approximations to the solution. Intuitively, we would expect piecewise parabolic curves to be a better approximation than piecewise linear ones, piecewise cubics still better and so forth.

Parabolic and higher degree basis functions lead to more complex calculations. For the piecewise linear basis functions, only one of the basis functions takes a nonzero value at each node, and that basis function is supported over two grid intervals, according to (2.19). For parabolic basis functions written in the most common fashion, there will be three basis functions with unit value at each node. They may be written as follows:

$$\phi_i^1, i = 1, \ldots = \begin{cases} \frac{1}{2h^2}[x-(i-2)h][x-(i-1)h], & (i-2)h \le x \le ih, \\ 0, & \text{otherwise}, \end{cases} \tag{2.26}$$

$$\phi_i^2, i = 1, \ldots = \begin{cases} -\frac{1}{h^2}[x-(i-1)h][x-(i+1)h], & (i-1)h \le x \le (i+1)h, \\ 0, & \text{otherwise}, \end{cases} \tag{2.27}$$

$$\phi_i^3, i = 1, \ldots = \begin{cases} \frac{1}{2h^2}[x-(i+2)h][x-(i+1)h], & ih \le x \le (i+2)h, \\ 0, & \text{otherwise}. \end{cases} \tag{2.28}$$

Clearly evaluation of the integrals in (2.20) for this basis set would be quite a bit more complex than the corresponding calculation in the example with piecewise linear basis functions. The problem becomes still more complex for nonlinear problems.

With this in mind, we might consider adjusting both the degree of the approximation space and the size of the intervals. By this procedure we arrive at the *spectral element methods.* Simple examples of application of spectral element methods were given by Patera (1984).

We may also take the domain as a whole as the only element, and construct our approximate solution from a space of functions with desirable properties such as Fourier components. This choice yields the *spectral methods.* The orthogonality of Fourier series implies that the system of ordinary differential equations satisfied by the coefficient functions will be fully explicit. A given scheme can, in theory, be made more accurate, at the cost of more work, by including more basis functions in the calculation. If we seek an approximate solution of the form (2.18), with $\phi_j = \mathrm{e}^{ik_j x}$, to the one-dimensional advection equation (2.1), the resulting semidiscrete scheme is simply

$$\dot{u}_j + ik_j u_j = 0. \tag{2.29}$$

Each Fourier component propagates at exactly the right speed, with no dissipation. The accuracy of the method is determined by the number of Fourier components included. For an initial condition such as the one given in Figure 2.1, a fairly large number of Fourier components would be necessary to represent the function faithfully because of the sharp corners, i.e., discontinuities in the derivative. For this simple example, the difference between a spectral method and every other method we have examined so far can be phrased in terms of fidelity to the true solution, component by component in Fourier space. The spectral method produces a perfect result for a fixed number of components and then sets all others to zero. All other methods track Fourier components of increasing wavenumber with decreasing accuracy, down to the component with wavelength $2\Delta x$.

In more than one space dimension, finite-element methods offer good accuracy as well as flexibility of treatment of domains with complex geometry, at a considerable cost in computational complexity. We will revisit this topic in subsequent sections. The standard engineering reference on finite-element methods is Zienkiewicz (1977). A more mathematically oriented, but still practical treatment can be found in Strang and Fix (1973). Allen *et al.* (1988) contains guidance for implementation of finite-element methods, as well as description of the algorithms. Irons and Shrive (1983) contains a more intuitive treatment aimed at the student of engineering.

It is in the case of problems governed by a variational principle that Galerkin methods really shine. In those cases, we are guaranteed, in some sense, the best approximation possible within the scope of the chosen basis functions. Consider, for example, the inhomogeneous Helmholtz equation

$$-u_{xx} + qu = f, \tag{2.30}$$

for $q(x) \geq 0$, with boundary conditions $u(0) = u'(\pi) = 0$. It is easily shown (see Exercise 2.8) that solving this equation is equivalent to finding the minimum of

$$J(u) = \int_0^\pi (u_x)^2 + qu^2 - 2fu\,\mathrm{d}x \qquad (2.31)$$
$$\equiv a(u,u) - \int_0^\pi 2fu\,\mathrm{d}x,$$

where we have defined the bilinear form $a(u,v) = \int_0^\pi u_x v_x + quv\,\mathrm{d}x$ for functions u, v that satisfy the boundary condition $u(0) = v(0) = 0$ and are smooth enough that the integral in (2.31) makes sense. Piecewise linear functions are not differentiable even once, so they cannot be considered as candidates for solution of (2.30) in the classical sense, but the integral in (2.31) is readily evaluated for piecewise linear u. For some choices of f, the minimizer of $J(u)$ may not be twice differentiable everywhere on $[0,\pi]$, and therefore is not strictly a solution of (2.30). The minimizer of $J(u)$ is said to be a *weak solution* of (2.30).

Suppose v is the minimizer of J over all admissible functions, and u^* is the minimizer of J over the space of piecewise linear functions, and therefore the solution by Galerkin's method of (2.30). It can be shown (see, e.g., Strang and Fix, 1973) that $a(u^* - v, u^* - v) \le a(v^* - v, v^* - v)$ for all admissible piecewise linear v^*. It can also be shown that this implies that $a(u,u)$ defines an equivalent norm to the L^2 norm $(\int_0^\pi u^2\,\mathrm{d}x)^{1/2}$, so u^* defines the solution that is "best" in the least-squares sense.

We must therefore have $a(u^* - v, u^* - v) \le a(\hat{u} - v, \hat{u} - v)$, where $\hat{u}$ is the linear interpolant of the true solution v, i.e., $\hat{u}$ is the piecewise linear function with $\hat{u}(x_i) = u(x_i)$ at the nodal points x_i; in other words, the Galerkin approximation is, in some sense, at least as good a solution as the linear interpolant of the true solution. The problem of estimating errors then reduces to approximation theory, in this case, the error bounds on linear interpolation (see Exercise 2.9). This is covered in Strang and Fix (1973) and Chapter 3 of Zienkiewicz (1977), among others.

2.6 Higher-order methods

So far, the methods we have considered have been second order, i.e., the errors decrease proportionately to the square of Δx or Δt. Generally, spatial accuracy has received more attention than temporal accuracy. Why?

Consider our practical tasks. Halving the time step, i.e., doubling the temporal resolution, increases our workload by a factor of two. We usually work in two or three space dimensions. Ocean basins have dimensions of thousands of kilometers, ten thousand kilometers or so for the Pacific. Typical dynamical scales are of the order of tens of kilometers, so basin scale models necessarily have huge numbers of gridpoints, even when resolution is severely compromised.

Dividing the horizontal grid spacing in half increases the number of gridpoints by four, and might increase the amount of work by an even larger factor due to the more stringent CFL condition imposed by the finer grid spacing. Halving the grid spacing in three dimensions increases the number of gridpoints by a factor of eight. If we consider the gain in accuracy, halving the grid spacing diminishes the error by a factor of four for a second-order method, and a factor of sixteen for a fourth-order method. Fourth-order methods involve more work, but the tradeoff may be favorable.

The easiest way to devise a fourth-order method is by *Richardson extrapolation*. Richardson extrapolation is the name given to procedures that increase the accuracy of a method by algebraically eliminating the leading error terms. Suppose we discretize the real line in increments of Δx so that $x_j = j\Delta x$, and consider the grid function $u_j = u(x_j)$. Then

$$\begin{aligned} u_{j+1} &= u_j + \Delta x u_{jx} + \frac{1}{2}\Delta x^2 u_{jxx} + \frac{1}{3!}\Delta x^3 u_{jxxx} + \frac{1}{4!}\Delta x^4 u_{jxxxx} \\ &\quad + O(\Delta x^5), \qquad (2.32) \\ u_{j-1} &= u_j - \Delta x u_{jx} + \frac{1}{2}\Delta x^2 u_{jxx} - \frac{1}{3!}\Delta x^3 u_{jxxx} + \frac{1}{4!}\Delta x^4 u_{jxxxx} \\ &\quad + O(\Delta x^5), \qquad (2.33) \end{aligned}$$

where u_{jx} is the partial derivative of u with respect to x at $x = x_j$. Then

$$\frac{u_{j+1} - u_{j-1}}{2\Delta x} = u_{jx} + \frac{1}{3!}\Delta x^2 u_{jxxx} + O(\Delta x^4) \qquad (2.34)$$

and

$$\frac{u_{j+2} - u_{j-2}}{4\Delta x} = u_{jx} + \frac{4}{3!}\Delta x^2 u_{jxxx} + O(\Delta x^4). \qquad (2.35)$$

Eliminating the lead error term and solving for u_{jx} yields our fourth-order method:

$$\frac{4}{3}\left(\frac{u_{j+1} - u_{j-1}}{2\Delta x}\right) - \frac{1}{3}\left(\frac{u_{j+2} - u_{j-2}}{4\Delta x}\right) = u_{jx} + O(\Delta x^4). \qquad (2.36)$$

This is about twice as much work – we have to calculate two difference quotients instead of one each time we calculate a derivative, so each halving of the grid step increases the accuracy by a factor of 16 while the work increases by a factor of only 8 for a problem in two space dimensions. For a second-order method in a two-dimensional problem, halving the grid step increases both accuracy and work by a factor of 4. From this point of view, fourth-order methods appear to have an advantage.

So why aren't fourth-order methods commonly used? One major problem is the question of the boundary. The method (2.36) requires values of the solution two gridpoints to either side of the point at which the derivative is being calculated. If that point lies on a boundary, two fictitious points outside the domain must be created, or an assumption about a derivative of at least second order must be made without compromising the accuracy of the method.

We can derive a compact scheme, i.e., one that requires only values at neighboring points, but we shall see that this advantage comes at a price. With the help of (2.32) and (2.33) we can see that

$$u_{jxxx} = \frac{u_{(j+1)x} - 2u_{jx} + u_{(j-1)x}}{\Delta x^2} + O(\Delta x^2), \tag{2.37}$$

so (2.34) can be written

$$\frac{u_{j+1} - u_{j-1}}{2\Delta x} = u_{jx} + \frac{\Delta x^2}{3!}\left[\frac{u_{(j+1)x} - 2u_{jx} + u_{(j-1)x}}{\Delta x^2} + O(\Delta x^2)\right] + O(\Delta x^4), \tag{2.38}$$

and therefore

$$\frac{u_{j+1} - u_{j-1}}{2\Delta x} = \frac{1}{6}u_{(j+1)x} + \frac{2}{3}u_{jx} + \frac{1}{6}u_{(j-1)x} + O(\Delta x^4). \tag{2.39}$$

So if we calculate the centered difference on the left-hand side of (2.39) at every point, we can get two additional orders of accuracy at the cost of solving the tridiagonal system defined by the right-hand side of (2.39).

It is interesting to note the similarity between (2.39) and (2.21), the semidiscrete finite-element approximation to the simple advection equation $u_t = -u_x$. The solutions to the corresponding semidiscrete approximation for our fourth-order scheme (2.39) will satisfy the same dispersion relation (2.22) as its finite-element counterpart.

2.7 A first look at boundary-value problems

In the examples given of methods for approximate solution of the simple advection equation (2.1) periodic boundary conditions were applied. In many, if not most, problems in ocean modeling this is not appropriate.

In even the simple case of (2.1), we can begin to see some of the issues that arise in numerical calculation of solutions to boundary-value problems.

Let us first consider the problem of finding solutions of (2.1) in the strip in the $x-t$ plane defined by $0 \leq x \leq 1$ and $t \geq 0$. From the analysis of (2.1) in Section 2.1 it is clear that the boundary at $x=0$ must be treated differently from the boundary at $x=1$. At the left-hand boundary $t=0$, the solution must be specified, since for any point $(\hat{x}, \hat{t})$ in the domain, the domain of dependence lies on the line $x=0$ if $\hat{t} > \hat{x}$.

On the other hand, we are not free to specify the solution at the right-hand boundary $x=1$, since the domain of dependence of points on the boundary extends into the domain, and arbitrary specification of $u(1,t)$ can lead to conflicts with initial conditions and conditions at the other boundary. Of course, one could specify the solution $u(1,t)$ at every point on $x=1$ and solve the equation backward in time to the segment $t=0,\ 0<x<1$ and the left-hand boundary $t>0,\ x=0$, but this is an unlikely analog of any real ocean modeling problem. The useful intuition comes from the description of the solution of the problem in the unbounded domain as translation of the initial condition to the right with unit speed. The solution must be specified at the left-hand boundary because information comes from the left, and should not be specified on the right-hand boundary because the solution at $x=1$ is determined by the solution in the interior of the domain at some previous time.

Sometimes the form of a numerical scheme may require the specification of a quantity that one is, in general, not free to specify. Suppose we were to attempt to use the leapfrog scheme to solve the equation (2.1) with the boundary condition $u(0,t)=g(t)$ and the initial condition $u(x,0)=f(x)$. Following earlier notation, write the approximate solution at $x=j\Delta x,\ t=n\Delta t$ as u_j^n, where $N\Delta x=1$. We could write $u_j^0=f(j\Delta x)$ and determine u_j^1 by an upwind scheme. The leapfrog scheme could be applied at the leftmost interior point $x=\Delta x$ by imposing the boundary condition where u_0^n was required, but what about u_{N-1}^n? A glance at the leapfrog formula (2.11) shows that a value is needed on the boundary $x=N\Delta x=1$. Specification of an arbitrary value on the boundary, say $u_N^n=0$, will ruin the entire calculation, since the incorrectly specified boundary condition will propagate into the interior by the leftward-propagating computational mode described in Section 2.4. This is worked out in some detail in Kreiss (1970).

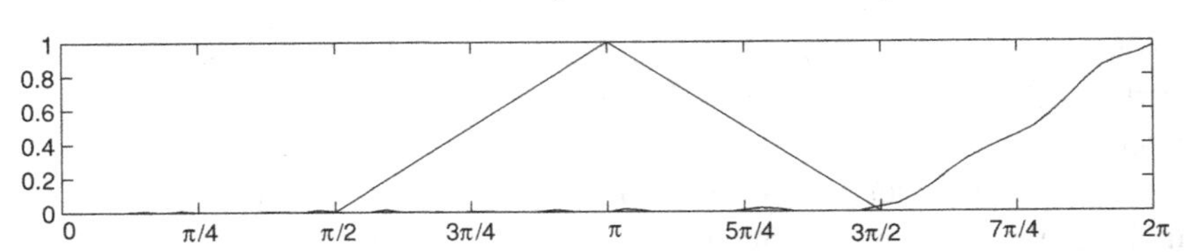

Fig. 2.13 Result of application of the Orlanski scheme to the simple advection equation (2.1). Initial condition, shown, is as in Figure 2.1. Curve sloping up toward the right-hand boundary is the approximate solution at $t = \pi$. Solution is by leapfrog, with $\mathrm{d}x = 2\pi/64$, $\lambda = 0.5$.

This simple example contains the seeds of the basic intuition for open boundary problems. The key is the nature of the characteristics, which were introduced in Section 2.1. In this case, if we consider integrations forward in time, the characteristics can be seen to carry information from left to right. This means that the information they carry must be specified at the left-hand boundary, and can only be specified on the right-hand boundary in terms of interior quantities.

Orlanski (1976) proposed an open-boundary scheme that forms the basis of many common open boundary methods. This scheme involves an explicit estimate of the phase speed of the outgoing waves.

To fix the ideas, let us choose the eastern boundary. For the state variable ϕ, write a local equation for the outgoing waves:

$$\phi_t + c\phi_x = 0.$$

Now calculate a local estimate of the phase speed c:

$$c = -\frac{\phi^k_{B-1} - \phi^{k-2}_{B-1}}{\left[(\phi^k_{B-1} + \phi^{k-2}_{B-1})/2 - \phi^{k-1}_{B-2}\right]} \frac{\Delta x}{2\Delta t}.$$

The solution to the local wave equation is then calculated by

$$\phi^{k+1}_B = \frac{[1 - c(\Delta t/\Delta x)]}{[1 + c(\Delta t/\Delta x)]}\phi^{k-1}_B + \frac{2c\Delta t/\Delta x}{[1 + c(\Delta t/\Delta x)]}\phi^k_{B-1}.$$

Application of this scheme to the simple advection equation (2.1) works quite well, as shown in Figure 2.13. The Orlanski scheme supposedly works well in a number of more relevant cases, but it is obviously difficult to implement. It also suffers from being an inherently nonlinear method; even in a linear model, this boundary scheme would render the whole problem nonlinear. This feature also makes the familiar linearized analyses all but impossible. For these reasons, this method should be viewed with caution.

2.8 Well- and ill-posed problems

In practice we expect that small changes in the inputs to a numerical scheme will result in small changes in the output. More precisely, we expect the output to be a continuous function of the input. This is similar to the definition of stability given in Section 2.2.1. In the language of calculus, for a problem with input u_0 output u_1, if we fix $\epsilon > 0$, then we can find $\delta > 0$ such that if the input is within δ of u_0, then the output will be within ϵ of u_1. A problem with this property is said to be *well-posed*; a problem that fails to have this property is said to be *ill-posed.* The question of whether a problem is well- or ill-posed depends on the choice of norm. A problem may be well-posed in one norm and ill-posed in another.

A classical example of an ill-posed problem is the backwards heat equation. Suppose we are given an initial temperature distribution u_0 at time $t = 0$, and asked to calculate an approximate solution at time $t = -1$. This problem has a unique solution: let u_{-1} be the solution at $t = -1$. Now consider an initial condition given by $u_0 + \delta \sin kx$. The solution at $t = -1$ is given by $u_{-1} + \delta e^{k^2} \sin kx$. Clearly, for any fixed ϵ, we can never pick δ such that $\delta e^{k^2} < \epsilon$ for all k. Put another way, we can find an arbitrarily small perturbation that will grow beyond any pre-set bound as time proceeds backwards from zero to -1.0, if we are only willing to make the perturbation wiggly enough.

It should be clear that ill-posed calculations are poor candidates for direct numerical solution. In the case of the backwards heat equation, no matter how smooth the initial condition is, there will always be some component of roundoff error which will introduce a wiggly perturbation. For a fixed choice of Δx, for fixed amplitude, the grid can only resolve so many wavenumbers, and fixed amplitude solutions are limited by resolution in their variability. Refining the grid will only make things worse, however, since the refined grid will admit solutions with greater spatial variability and therefore more rapid growth as the computation proceeds backwards in time. Much has been written about numerical solution of ill-posed problems, but the reader should be reminded that this is a matter of terminology. As a practical matter, one can never solve an ill-posed problem numerically. When people say they have solved an ill-posed problem, they have, in fact, replaced the ill-posed problem with a well-posed problem, solved the well-posed problem and then argued that the solution so obtained is relevant to the original motivation from which the ill-posed problem was derived. We will encounter ill-posed problems when we examine open boundary problems.

2.9 Exercises

2.1 Consider the simple linear advection equation $u_t + u_x = 0$. The *downwind scheme* is given by

$$u_j^{n+1} = u_j^n - \lambda(u_{j+1}^n - u_j^n),$$

where $\lambda = \Delta t/\Delta x$. Show that the downwind scheme fails to satisfy the CFL criterion and is therefore unstable.

2.2 Find the CFL condition for the simple unstable centered scheme given in (2.3).

2.3 Consider the 1-D linearized barotropic quasigeostrophic potential vorticity equation

$$\psi_{xxt} + U\psi_{xxx} + \psi_x = 0. \qquad \text{(E2.1)}$$

(i) Find wavelike solutions. These are the *Rossby waves.*

(ii) Show that for the initial value problem with initial conditions which vanish outside of a bounded interval (i.e., the initial function has *compact support*) the quantity

$$\frac{1}{2}\int_{-\infty}^{\infty} \psi_x^2 \, \mathrm{d}x$$

is conserved.

(iii) Consider the semidiscrete approximation to (E2.1), where only spatial derivatives are approximated by differences. For $U > 0$, an upstream differencing scheme would be

$$\begin{aligned}\dot{\psi}_{j+1} - 2\dot{\psi}_j + \dot{\psi}_{j-1} \\ = -\frac{U}{\Delta x}(\psi_{j+1} - 3\psi_j + 3\psi_{j-1} - \psi_{j-2}) - \Delta x(\psi_j - \psi_{j-1}).\end{aligned}$$

ψ_j is now a function of time; raised dots denote time differentiation. Look for solutions of the form

$$\psi_j = \exp\left[i(jk\Delta x - \omega t)\right].$$

Show that the discrete dispersion relation is

$$\omega = -ic_{\Delta x}(k)\left(\frac{1-\cos k\Delta x}{\Delta x}\right) + c_{\Delta x}(k)\left(\frac{\sin k\Delta x}{\Delta x}\right),$$

where $c_{\Delta x} = U + \Delta x^2/2(\cos k\Delta x - 1)$.

Show that $c_{\Delta x}(k)$ approaches the true phase speed $c(k)$ of the Rossby waves as $\Delta x \to 0$, and that for sufficiently small k, the semidiscrete solutions will always grow exponentially in time.

(iv) Construct a similar scheme for (E2.1) in the case $U < 0$. Again, use the approximation

$$\psi_{xx} \approx \frac{\psi_{j+1} - 2\psi_j + \psi_{j-1}}{\Delta x^2}$$

and use upstream differencing for the advection term, and for the "ψ_x" term in (E2.1). Show that, in this case, solutions always decay in time. For reference, see, e.g., Miller (1986).

2.4 Recall that the leapfrog scheme

$$u_j^{n+1} = u_j^{n-1} - \lambda(u_{j+1}^n - u_{j-1}^n)$$

requires values for u_j^1 as well as u_j^0 in order to initiate computation. In order to obtain u_j^1, one might suggest using the scheme

$$u_j^1 = u_j^0 + \frac{\lambda}{2}(u_{j+1}^0 - u_{j-1}^0).$$

Since we know this scheme to be unstable, is this a reasonable thing to do? Why or why not?

2.5 (i) Show that the scheme

$$\begin{aligned} u_j^{n+1} &= u_j^n + v_{j+1}^n - 2v_j^n + v_{j-1}^n \\ v_j^{n+1} &= v_j^n \end{aligned}$$

is consistent with the differential equation

$$\frac{\partial}{\partial t}\begin{pmatrix} u \\ v \end{pmatrix} = 0.$$

(Note that $v_{j+1}^n - 2v_j^n + v_{j-1}^n$ approaches $\Delta x^2 v_{xx}$ as $\Delta x \to 0$.)

(ii) Show that this scheme satisfies the von Neumann condition.

(iii) Show that this scheme is unstable.

Hint 1: refer to your notes on the leapfrog scheme with $\lambda = 1$.
Hint 2: the identity $\sin^2(\Theta/2) = (1 - \cos\Theta)/2$ may be convenient.

2.6 Consider the *heat equation*

$$u_t = u_{xx}$$

for initial conditions with compact support on the real line.

(i) Show that, for any solution u, the quantity $\int_{-\infty}^{+\infty} u^2 dx$ cannot increase with time. Given this fact, do you think that the definition of stability of a numerical scheme given in class is a good one to apply to methods for the heat equation? If not, can you think of a more useful definition of stability?

(ii) Investigate the stability of the difference scheme

$$u_j^{n+1} = u_j^n + \frac{\Delta t}{\Delta x^2}(u_{j+1}^n - 2u_j^n + u_{j-1}^n).$$

(iii) Show that the leapfrog scheme for the heat equation

$$u^{n+1} = u^{n-1} + \frac{2\Delta t}{\Delta x^2}(u_{j+1}^n - 2u_j^n + u_{j-1}^n)$$

is unstable.

2.7 The *sup norm* of a vector $\mathbf{v}$ is the greatest of the absolute values of its components. Characterize the matrix norm induced by the sup norm in terms of the components of the matrix. You may find it easiest to begin with 2×2 matrices and generalize from there.

2.8 *Equation derived from a variational principle:* Apply the calculus of variations to show that the minimizer u of the functional $J(u)$ defined in (2.31) is a solution to (2.30) with the given boundary conditions.

2.9 *Linear interpolation:* Let $\hat{u}$ be the linear interpolant of a twice-differentiable function u on $[0, \pi]$. Show that $\| \hat{u} - u \|_2 \leq Ch^2 |u''|^2$ for grid spacing h and some constant C.

3

Shallow-water models: the simplest ocean models

3.1 Introduction

What do we want to model? Among other things, we want to model the "general circulation": basin (and larger) scale wind and thermally driven circulation. We also wish to model eddies, jets and other intense features.

I'll bet you think the ocean is deep. The ocean is shallow. A typical ocean basin has about the proportions of an 8.5×11 inch sheet of paper. So while it may be deep compared to your height or to the depth to which you can safely dive, the ocean is shallow compared to its horizontal extent, and can be safely considered physically for many purposes as a shallow fluid. It might therefore seem appropriate to write down the equations of motion for a shallow fluid, that is, one in which the vertical scales of motion are negligible compared to the horizontal ones. These are the shallow-water equations. We write them here in Cartesian coordinates in linearized form:

$$u_t + gh_x - fv = 0, \tag{3.1}$$

$$v_t + gh_y + fu = 0, \tag{3.2}$$

$$h_t + H(u_x + v_y) = 0. \tag{3.3}$$

Here u and v are the velocity components in the x and y directions respectively, h is the deviation of the depth of the water from its resting value H and g is the acceleration of gravity. The Coriolis parameter f

is assumed constant for now. We can write this as

$$\begin{pmatrix} u \\ v \\ h \end{pmatrix}_t + \begin{pmatrix} 0 & 0 & g \\ 0 & 0 & 0 \\ H & 0 & 0 \end{pmatrix} \begin{pmatrix} u \\ v \\ h \end{pmatrix}_x + \begin{pmatrix} 0 & 0 & 0 \\ 0 & 0 & g \\ 0 & H & 0 \end{pmatrix} \begin{pmatrix} u \\ v \\ h \end{pmatrix}_y$$
$$+ \begin{pmatrix} 0 & -f & 0 \\ f & 0 & 0 \\ 0 & 0 & 0 \end{pmatrix} \begin{pmatrix} u \\ v \\ h \end{pmatrix} = 0.$$

Now look for solutions of the form

$$\begin{pmatrix} u \\ v \\ h \end{pmatrix} = \mathbf{V}_0 \mathrm{e}^{\mathrm{i}(\mathbf{k}\cdot\mathbf{x}-\omega t)}\,; \qquad \mathbf{V}_0 = \begin{pmatrix} u_0 \\ v_0 \\ h_0 \end{pmatrix}; \qquad \mathbf{k} = (k, l),$$

where $\mathbf{V}_0$ is a constant vector with components u_0, v_0 and h_0. We find that

$$-\mathrm{i}\omega \mathbf{V}_0 + \begin{pmatrix} 0 & -f & \mathrm{i}kg \\ f & 0 & \mathrm{i}lg \\ \mathrm{i}Hk & \mathrm{i}Hl & 0 \end{pmatrix} \mathbf{V}_0 = 0.$$

Therefore, ω is an eigenvalue of the matrix

$$\begin{pmatrix} 0 & \mathrm{i}f & kg \\ -\mathrm{i}f & 0 & lg \\ Hk & Hl & 0 \end{pmatrix}, \tag{3.4}$$

i.e., ω is a solution to

$$-\omega^3 + \omega[gH(k^2 + l^2) + f^2] = 0,$$

so $\omega = 0$ or $\omega = \pm\sqrt{gH(k^2 + l^2) + f^2}$.

We *really* don't like this. What will be the wave speeds?

$$C_{\mathrm{phase}} = \frac{\omega}{(k^2 + l^2)^{1/2}} = \pm\sqrt{gH + f^2/(k^2 + l^2)},$$

$$g \sim 10\,\mathrm{m\,s^{-2}}, \qquad H \sim 5000\,\mathrm{m}, \qquad gH \sim 5 \times 10^4\,\mathrm{m^2\,s^{-2}}.$$

So typical phase speeds C_{phase} will be $\sim 225\,\mathrm{m\,s^{-1}}$. These can't be the creatures we're interested in. Computing them would be extremely demanding; consider the CFL condition for a reasonable calculation. These are the *surface gravity waves*. Undisturbed, they would cross a major ocean basin in a matter of hours. They evolve on timescales much faster than those normally considered in ocean circulation modeling.

What about the $\omega = 0$ solution? The relevant eigenvector is

$$\begin{pmatrix} u_0 \\ v_0 \\ h_0 \end{pmatrix} = \begin{pmatrix} -\mathrm{i}lg/f \\ \mathrm{i}kg/f \\ 1 \end{pmatrix}.$$

This is the steady geostrophically balanced state

$$fu = -gh_y,$$
$$fv = gh_x.$$

We need a better model. How shall we make one? Think again about the physics: we didn't really want shallow-water motion. This is bulk motion. The motions of interest will be significantly affected by stratification. Let us therefore try modeling the stratified ocean as two immiscible fluids of different densities ρ_1 and ρ_2 in the upper and lower layers respectively. We assume the system to be statically stable, i.e., $\rho_2 > \rho_1$. The speeds u_1 and u_2 in the upper and lower layers are assumed to be vertically uniform. The thicknesses of the two layers are given by h_1 and h_2, both of which are functions of space and time. The undisturbed thicknesses of the two layers are given by H_1 and H_2. Our two-layer model is illustrated in Figure 3.1.

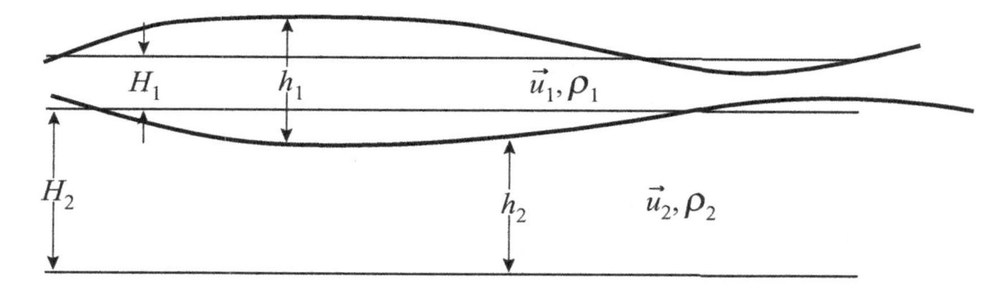

Fig. 3.1 Schematic diagram of a simple two-layer model.

We assume free slip at the interface, so the two layers are coupled only through the pressure term. For simplicity, we examine the case of one horizontal space dimension without rotation, and we neglect nonlinear terms. The physical situation pictured in Figure 3.1 is described by

$$u_{1t} + g(h_1 + h_2)_x = 0, \tag{3.5}$$

$$h_{1t} + H_1 u_{1x} = 0, \tag{3.6}$$

$$u_{2t} + g(\rho_1/\rho_2)h_{1x} + gh_{2x} = 0, \tag{3.7}$$

$$h_{2t} + H_2 u_{2x} = 0. \tag{3.8}$$

Now look for wavelike solutions, i.e., solutions of the form

$$\begin{pmatrix} u_1 \\ h_1 \\ u_2 \\ h_2 \end{pmatrix} = \mathbf{V}_0 \mathrm{e}^{\mathrm{i}(kx-\omega t)}.$$

We find that

$$-\mathrm{i}\omega \begin{pmatrix} u_1 \\ h_1 \\ u_2 \\ h_2 \end{pmatrix} + \mathrm{i} \begin{pmatrix} 0 & kg & 0 & kg \\ kH_1 & 0 & 0 & 0 \\ 0 & kg\rho_1/\rho_2 & 0 & kg \\ 0 & 0 & kH_2 & 0 \end{pmatrix} \begin{pmatrix} u_1 \\ h_1 \\ u_2 \\ h_2 \end{pmatrix} = 0, \qquad (3.9)$$

so ω is an eigenvalue of the matrix on the right, i.e.,

$$\det \begin{pmatrix} -\omega & kg & 0 & kg \\ kH_1 & -\omega & 0 & 0 \\ 0 & gk\rho_1/\rho_2 & -\omega & gk \\ 0 & 0 & kH_2 & -\omega \end{pmatrix} = 0,$$

or

$$-\omega \det \begin{pmatrix} -\omega & 0 & 0 \\ gk\rho_1/\rho_2 & -\omega & gk \\ 0 & kH_2 & -\omega \end{pmatrix} - kg \det \begin{pmatrix} kH_1 & 0 & 0 \\ 0 & -\omega & gk \\ 0 & kH_2 & -\omega \end{pmatrix}$$

$$-kg \det \begin{pmatrix} kH_1 & -\omega & 0 \\ 0 & gk\rho_1/\rho_2 & -\omega \\ 0 & 0 & kH_2 \end{pmatrix}$$

$$\begin{aligned} &= \omega^2[\omega^2 - k^2 gH_2] - gk^2 H_1(\omega^2 - k^2 gH_2) - gk[k^3 H_1 H_2 g\rho_1/\rho_2] \\ &= \omega^4 - \omega^2 k^2 g(H_1 + H_2) + g^2 k^4 H_1 H_2 (1 - \rho_1/\rho_2) \\ &= 0. \end{aligned}$$

ω^2 is therefore the solution of a quadratic equation:

$$\omega^2 = \frac{1}{2} k^2 g(H_1 + H_2) \left[1 \pm \sqrt{1 - 4 \left(\frac{\Delta\rho}{\rho_2} \right) \left(\frac{H_1 H_2}{(H_1 + H_2)^2} \right)} \right],$$

where $\Delta\rho = \rho_2 - \rho_1$. Now $\Delta\rho/\rho_2$ is small (typically ≈ 0.002), and $(H_1 + H_2)^2/4 - H_1 H_2 = (H_1 - H_2)^2/4 \geq 0$ (i.e., the arithmetic mean is always greater than the geometric mean), hence $H_1 H_2/(H_1 + H_2)^2 \leq 1$ and we may approximate ω as

$$\omega^2 \approx \frac{1}{2} k^2 g(H_1 + H_2) \left\{ 1 \pm \left[1 - 2 \left(\frac{\Delta\rho}{\rho_2} \right) \left(\frac{H_1 H_2}{(H_1 + H_2)^2} \right) \right] \right\}. \qquad (3.10)$$

Therefore the "+" root of (3.10) leads to

$$C^2_{\text{phase}} \equiv \frac{\omega^2}{k^2} = g(H_1 + H_2)\left[1 - \left(\frac{\Delta\rho}{\rho_2}\right)\left(\frac{H_1 H_2}{(H_1 + H_2)^2}\right)\right], \tag{3.11}$$

while the "−" root leads to

$$C^2_{\text{phase}} = g\left(\frac{\Delta\rho}{\rho_2}\right)\left(\frac{H_1 H_2}{H_1 + H_2}\right). \tag{3.12}$$

From (3.11) we see that the faster of the two travels at approximately the shallow-water wave speed for a fluid with depth $H_1 + H_2$. For $H_2 >> H_1, H_1 H_2/(H_1 + H_2) \approx H_1$ and the speed (3.12) of the slower wave is well approximated by $C^2_{\text{phase}} \approx g(\Delta\rho/\rho_2)H_1$. This is what the ordinary shallow-water wave speed would be for a fluid of depth H_1 and gravitational acceleration equal to $g\Delta\rho/\rho_2$. The quantity $g\Delta\rho/\rho_2 = g'$ is known as the *reduced gravity.*

The relation between the speeds u_1 and u_2 in the upper and lower layers is determined by the eigenvectors of the matrix in (3.9). From (3.9) we derive

$$gk^2(H_1 u_1 + H_2 u_2) = \omega^2 u_1, \tag{3.13}$$

$$gk^2\left[\left(\frac{\rho_1}{\rho_2}\right) H_1 u_1 + H_2 u_2\right] = \omega^2 u_2. \tag{3.14}$$

For the faster of the two waves, i.e., the one with speed given by (3.11), we can simply divide both sides of (3.13) by u_1 and substitute the form of ω^2 from the "+" root of (3.10) to find that the ratio of the lower layer speed to the upper layer speed is given approximately by

$$\frac{u_2}{u_1} = 1 - \frac{(\Delta\rho/\rho_2)H_1}{H_1 + H_2},$$

so the speed in the lower layer is slightly slower than that in the upper layer, and directed in the same direction. As $H_1/H_2 \to 0$ and/or $\Delta\rho/\rho_2 \to 0$, we recover ordinary shallow-water motion. This mode is not strongly affected by the stratification. Because of the presence of this mode, we may say that for the stratifications we are likely to encounter, the stratified system admits a mode that is essentially identical to the bulk motions exhibited by the unstratified system.

For the slower of the two waves, we combine (3.13) and (3.14) to find

$$\omega^2(u_1 - u_2) = k^2 g H_1\left(\frac{\Delta\rho}{\rho_2}\right) u_1. \tag{3.15}$$

Substitution of the "–" root of (3.10) leads to

$$\left[1 - \frac{u_2}{u_1}\right]\left(\frac{\Delta\rho}{\rho_2}\right)\left(\frac{H_2}{H_1 + H_2}\right) = \frac{\Delta\rho}{\rho_2}, \tag{3.16}$$

so

$$\frac{u_2}{u_1} \approx -\frac{H_1}{H_2}. \tag{3.17}$$

For $H_1 << H_2$, the velocity in the lower layer relative to that in the upper layer is very small in magnitude and directed in the opposite direction to that of the upper layer. What about the limit $\Delta\rho/\rho_2 \to 0$? We note that in order to obtain (3.17) we had to cancel $\Delta\rho/\rho_2$ on both sides of (3.16). Examination of (3.15) leads to the expected conclusion that $u_2 \to u_1$ as $\Delta\rho/\rho_2 \to 0$.

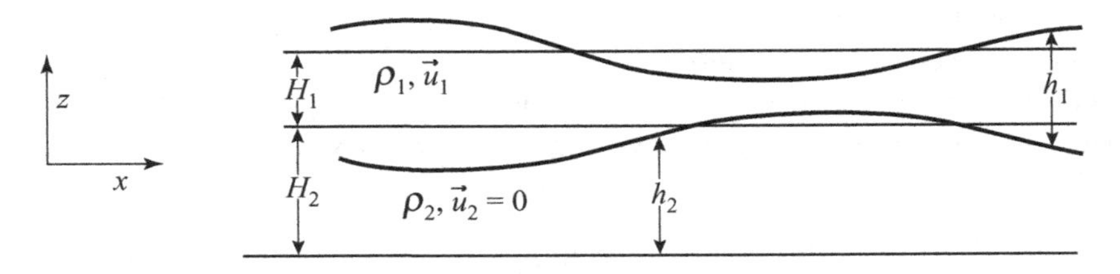

Fig. 3.2 Schematic diagram of the reduced-gravity model. Similar to Figure 3.1 but velocity is assumed to vanish in the lower layer.

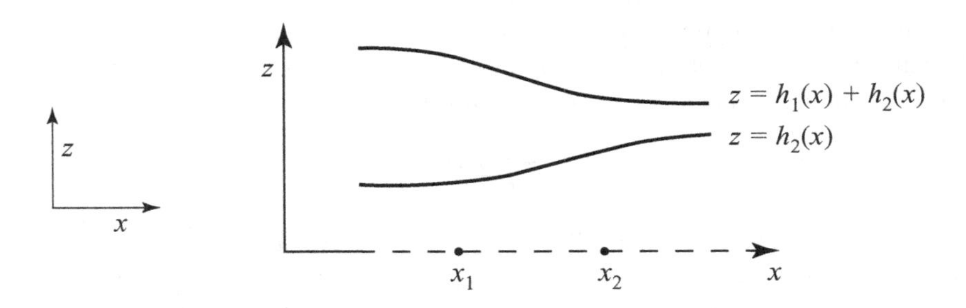

Fig. 3.3 Illustration of calculation of pressure gradient in the lower layer of our reduced-gravity model.

This slower wave mode evolves on the relevant scales and is therefore the one we are interested in, but how can we keep the fast wave out of our approximate solution? If we simply implement the two-layer model, the fast waves will appear, whether we want them or not, and these fast waves will determine the CFL conditions for our calculations.

The simplest thing we can do is derive the reduced-gravity formulation directly by assuming that the velocity in the lower layer vanishes. Assume $H_2 >> H_1$ and $u_2 \equiv 0$, as shown in Figure 3.2. How shall we arrange this? If the velocity in the lower layer is to vanish for all time, then pressure gradients must vanish in the lower layer also. We can now derive the condition that the pressure gradient must vanish in the lower layer (see Figure 3.3):

$$\rho_1 h_1(x_1) + \rho_2 h_2(x_1) = \rho_1 h_1(x_2) + \rho_2 h_2(x_2).$$

Rearranging terms and dividing by $x_2 - x_1$ leads to

$$\rho_1 h_{1x} = -\rho_2 h_{2x}.$$

h_2 can then be eliminated from the momentum equation (3.5) for the upper layer, and the motion in the upper layer can then be described simply by the *reduced-gravity equations*:

$$u_{1t} + g' h_{1x} = 0, \tag{3.18}$$

$$h_{1t} + H_1 u_{1x} = 0. \tag{3.19}$$

One thing we can do is solve (3.18)–(3.19). This eliminates the surface waves and the stringent CFL condition that goes with them, but the reduced-gravity approximation is adequate only for rough qualitative descriptions in most ocean regimes. Usually $u_2 = 0$ is not a very good approximation and the reduced-gravity approximation can not allow for variations in total depth. We shall see other ways to formulate computationally tractable models later.

3.2 Discrete methods for the shallow-water equations

We begin this section with the simplest case, i.e., the shallow-water equations in one space dimension without rotation:

$$\begin{aligned} u_t + g h_x &= 0, \\ h_t + H u_x &= 0. \end{aligned}$$

If we first consider a semidiscrete approximation of the shallow-water equations along the x-axis:

$$\begin{aligned} \dot{u}_j &= -\frac{g}{2\Delta x}(h_{j+1} - h_{j-1}), \\ \dot{h}_j &= -\frac{H}{2\Delta x}(u_{j+1} - u_{j-1}), \end{aligned}$$

we find that u depends on h at adjacent points, but not on u at adjacent points, and similarly for h. We thus find a natural checkerboard effect arising from the form of the equations. In the interests of economy, we may therefore choose to compute only one of the two independent solutions. We can do this by using a *staggered grid*, i.e., u is evaluated at points with even indices, and h is evaluated at odd points, or vice versa.

We can perform the CFL analysis for the staggered-grid system with time discretization by the leapfrog scheme. The relevant domains of dependence are depicted in Figure 3.4. It is clear from Figure 3.4 that the necessary condition for stability is dictated by the spacing $\Delta x/2$ between adjacent points, rather than the spacing Δx between points at which like quantities are evaluated. The example shown in Figure 3.4 satisfies the CFL condition.

In two space dimensions, we have several choices of grids. Four choices, along with the names by which they are known in the literature, are depicted in Figure 3.5; the nomenclature dates back to the 1970s; see, e.g., Mesinger and Arakawa (1976) or Arakawa and Lamb (1977). Consider the form of the zonal momentum equation. For the C-grid, if the zonal acceleration is evaluated at one of the u points shown in Figure 3.5(c), then the pressure gradient is evaluated as a centered difference of the adjacent pressure points, and the Coriolis term is evaluated as an average of the surrounding v points. The difference quotients in the mass conservation equation are also centered in a natural way. For the D-grid, geostrophic balance is easily expressed in terms of simple centered expressions, but the balance between acceleration and pressure gradient terms cannot be expressed so simply in the momentum equations. A similar remark applies to the mass equation.

In the reduced-gravity model, we may scale the depth h by the resting depth H of the active layer, the speeds u and v by the gravity wave speed $c = (g'H)^{1/2}$ and the lengths x and y by the internal deformation radius c/f. With this scaling, we may write the semidiscrete approximation to the shallow-water equations (3.1)–(3.3) on the B-grid with grid spacing d as

$$\begin{aligned}
\dot{u}_{j,k} - v_{j,k} &= -\frac{1}{d}[h_{j+1/2,k+1/2} - h_{j-1/2,k+1/2} \\
&\qquad + h_{j-1/2,k+1/2} - h_{j-1/2,k-1/2}], \\
\dot{v}_{j,k} + u_{j,k} &= -\frac{1}{d}[h_{j+1/2,k+1/2} - h_{j+1/2,k-1/2} \\
&\qquad + h_{j+1/2,k-1/2} - h_{j-1/2,k-1/2}],
\end{aligned}$$

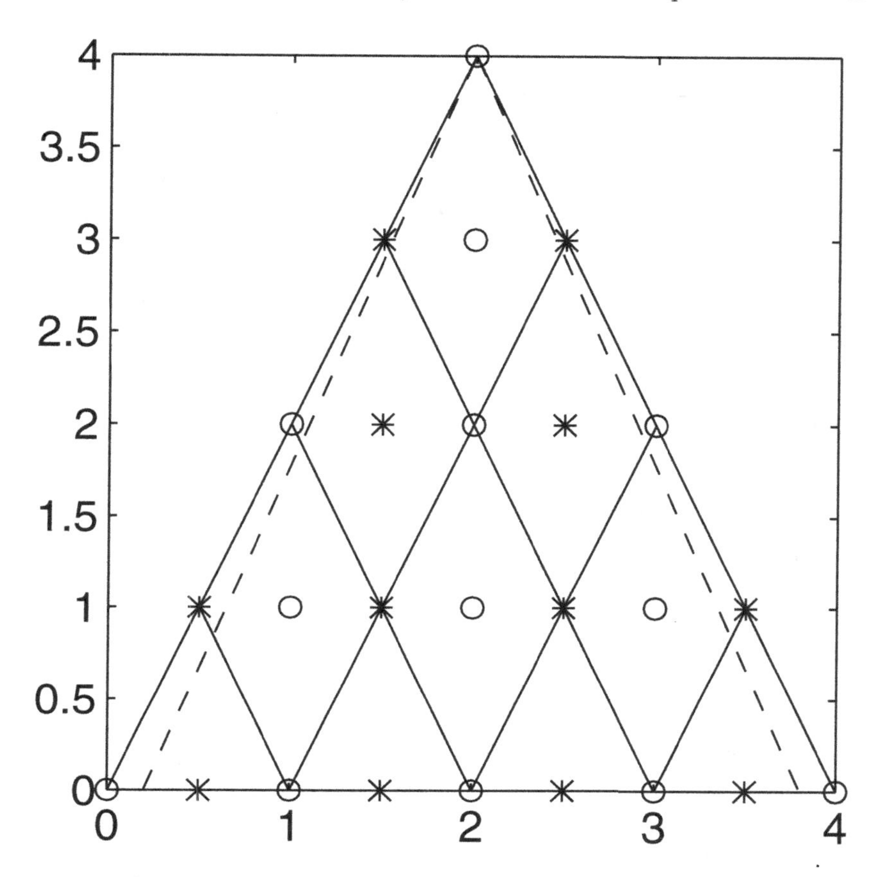

Fig. 3.4 Diagram in $x - t$ space of the staggered-grid system for the shallow-water equations with time discretization by the leapfrog scheme. The speed u is calculated at the points shown as open circles and the pressure is calculated at the points marked by $*$. The solid lines depict the boundaries of computational domains of influence. The dashed line has slope $1/(gH)^{1/2}$; it defines the domain of influence for the true system.

$$\dot{h}_{j+1/2,k+1/2} = -\frac{1}{d}[u_{j+1,k+1} - u_{j,k+1} + u_{j+1,k} - u_{j,k}] \\ - \frac{1}{d}[v_{j+1,k+1} - v_{j+1,k} + v_{j,k+1} - v_{j,k}].$$

Applying the discrete Fourier transform, with wavenumbers μ and ν in the x- and y-directions respectively, yields

$$\dot{\hat{u}} - \hat{v} = -\frac{i}{d}[\sin(\mu d)\cos(\nu d)]\hat{h},$$

$$\dot{\hat{v}} + \hat{u} = -\frac{i}{d}[\sin(\nu d)\cos(\mu d)]\hat{h},$$

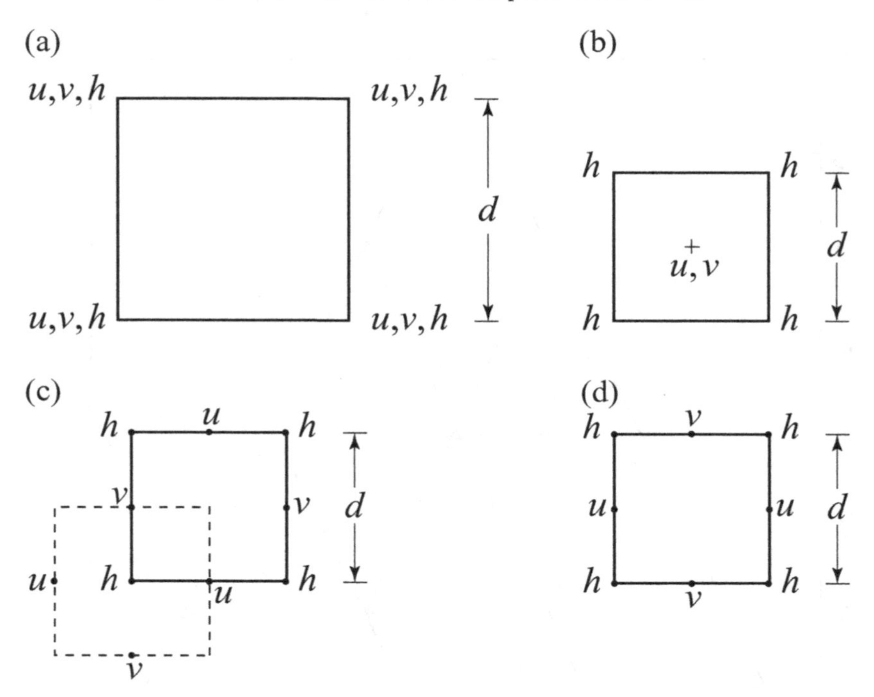

Fig. 3.5 Schematic diagrams of common choices of computational grids. (a) In the A-grid, pressure and both horizontal velocity components are evaluated at every point. (b) The B-grid: in the grid square shown with pressure evaluated at the corners, the horizontal velocity components are evaluated at the center; this pattern repeats in such a way that in grid squares with the velocity components at the corners, the pressure is evaluated at the center. (c) The C-grid: in grid squares with pressure evaluated at the corners, the zonal velocity component u is evaluated at the midpoints of the upper and lower edges, while the meridional velocity component v is evaluated at the midpoints of the left and right edges. (d) The D-grid is similar to the C-grid, but with v evaluated at the upper and lower edges and u evaluated at the left- and right-hand edges.

$$\dot{\hat{h}} = -\frac{i}{d}[\sin(\mu d)\cos(\nu d)\hat{u} + \sin(\nu d)\cos(\mu d)\hat{v}].$$

The frequencies of wavelike solutions to these equations will be given by the eigenvalues of a matrix similar to (3.4), with k replaced by $\sin(\mu d)\cos(\nu d)/d$ and l replaced by $\sin(\nu d)\cos(\mu d)/d$. Similar analyses can be performed for the A- and C-grids. For $\nu = 0$, i.e., motion independent of y, the dispersion relations are shown in Figures 3.6 and 3.7. These dispersion relations are worth examining in detail.

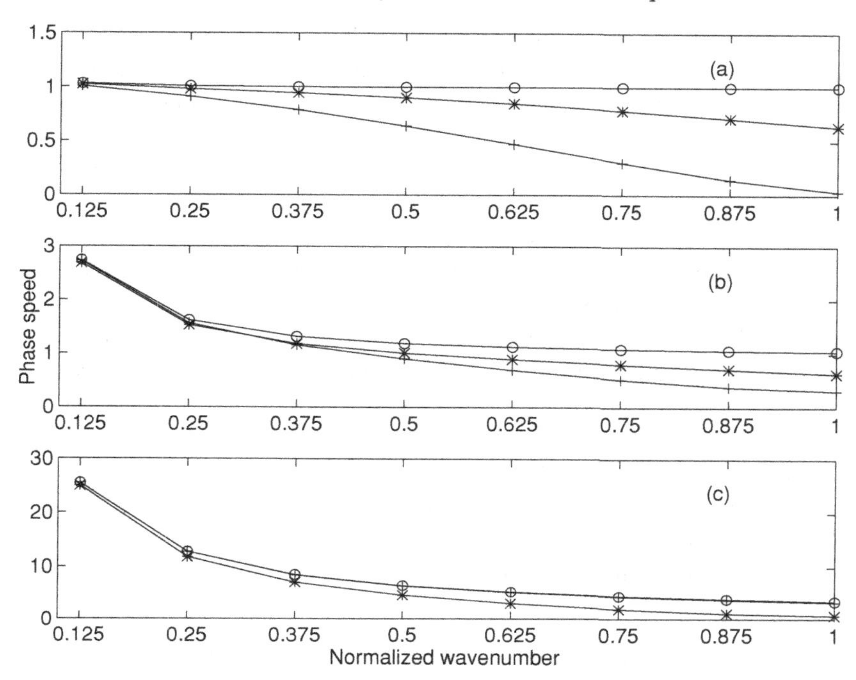

Fig. 3.6 Comparison of exact dispersion relation for the scaled shallow-water equations with dispersion relations derived from the semidiscrete approximations on the Arakawa B- and C-grids. Open circles denote the exact solution; stars "*" denote the C-grid; plus signs "+" denote the B-grid. Wavenumbers are normalized by π/d, where d is the grid spacing. (a) $d = 0.1$, (b) $d = 1$, (c) $d = 10$.

Figure 3.6 shows a comparison of the phase speeds of the wavelike solutions to the full differential equations and to the approximate equations on the B and C grids for different choices of the grid spacing d. Recall the scaling of these equations: lengths are scaled by the internal radius of deformation; a typical value for the North Atlantic would be 50 km. In each case, the phase speed is plotted against the normalized wavenumber kd/π, so the range from 0 to 1 represents the full range of resolvable wavenumbers. With the wavenumbers scaled by π/d, the highest resolved wavenumbers for the cases shown are 10π, π and 0.1π for panels (a), (b) and (c) respectively, so if the wavenumbers were not scaled, panel (b) would end where panel (a) begins, and panel (c) would end where panel (b) begins.

For the fine and intermediate values of the grid spacing d, the C-grid is superior, while the B-grid is better in the coarsely resolved case $d = 10$.

The reader should be reminded not to conclude from Figure 3.6(c) that the B-grid is superior overall because the phase speeds are almost exactly correct, since the range of resolved wavenumbers is very small in this case.

Comparison of the same information plotted slightly differently in Figure 3.7 shows further that, in the case of very coarse resolution, the group speed $c_g = d\omega/dk$ takes the wrong sign for the C-grid, even for small wavenumbers; see the lowest of the four curves in Figure 3.7(b). The exact dispersion relation is not displayed in Figure 3.7 because there would be a different exact solution curve for each approximate curve, so these curves were omitted to reduce clutter. The exact dispersion curves are given by $\omega = (1 + (k\pi/d)^2)^{1/2}$, where ω is the frequency and k is the normalized wavenumber.

The information contained in these graphs forms the basis for the usual convention of using the B-grid for coarsely resolved models and the C-grid for finely resolved models. More details and different presentations can be found in Wajsowicz (1986) and Blayo (2000).

3.3 Special considerations for nonlinear problems

3.3.1 Stability considerations for advection-diffusion equations

Up to this point we have written the equations of motion with no dissipation. Dissipation is important in the real ocean, and some dissipation is usually included in most ocean models, either for computational stability, or to balance external forcing. Most ocean models fall into a category often called "forced-dissipative systems." Advection is also important in real ocean models, so, like many fluid dynamical systems, ocean models contain advection and dissipation terms which can be comparable in magnitude. For this reason, most ocean model equations, like the Navier–Stokes equations, are called "advection-diffusion equations." The result of inadequate resolution in an advection-diffusion equation can be non-physical oscillations in the solution. This can be seen from consideration of the simple steady advection-diffusion equation,

$$uT_x - \nu T_{xx} = 0.$$

If we impose the boundary conditions

$$\begin{aligned} T(0) &= 0, \\ T(1) &= 1, \end{aligned}$$

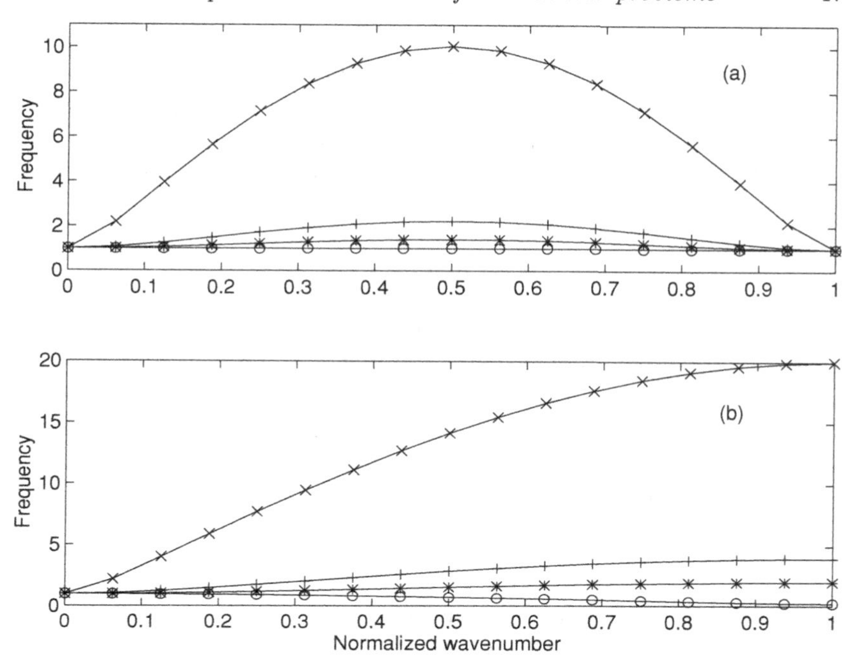

Fig. 3.7 (a) Dispersion diagrams for B-grid semidiscrete approximations to the shallow-water equations, plotted as frequency vs. wavenumber, for various choices of resolution d. Crosses "x" denote $d = 0.1$; plus signs "+" denote $d = 0.5$; stars "*" denote $d = 1$ and open circles "o" denote $d = 10$. (b) Similar to (a), but for the C-grid.

we find the exact solution to be

$$T = \frac{e^{ux/\nu} - 1}{e^{u/\nu} - 1}.$$

A consistent centered difference approximation to the given equation is

$$\frac{u}{2\Delta x}(T_{j+1} - T_{j-1}) - \frac{\nu}{\Delta x^2}(T_{j+1} - 2T_j + T_{j-1}) = 0,$$

which can be rearranged to form

$$-(1 + 0.5R_{\text{cell}})T_{j-1} + 2T_j - (1 - 0.5R_{\text{cell}})T_{j+1} = 0,$$

where R_{cell} is the *cell Reynolds number* $u\Delta x/\nu$. The solution to this difference equation is

$$T_j = A + B\left[\frac{1 + 0.5R_{\text{cell}}}{1 - 0.5R_{\text{cell}}}\right]^j,$$

where A and B are chosen to satisfy the boundary conditions. (The sharp-eyed among you with sharp memories will recognize the expression in brackets as a Padé approximant of the exponential; see, e.g., Ralston (1965), Ch. 7.) If $R_{\text{cell}} > 2$, the expression in brackets will be negative, and the solution will oscillate. This example was taken from Fletcher (1991); see also Weaver and Sarachik (1990).

3.3.2 Nonlinear instability

There are cases of nonlinear systems in which a scheme that is stable for the linearized system is unstable for the nonlinear problem. An example may be found in Richtmyer (1963). This behavior is usually attributed to *aliasing*. In this section, we will examine Richtmyer's example in detail.

Consider the model equation

$$u_t = uu_x = \left(\frac{1}{2}u^2\right)_x, \tag{3.20}$$

along with the leapfrog difference scheme

$$u_j^{n+1} - u_j^{n-1} = \frac{\Delta t}{2\Delta x}[(u_{j+1}^n)^2 - (u_{j-1}^n)^2]. \tag{3.21}$$

If we look for solutions to (3.20) of the form

$$u = A(t)\cos kx + B \tag{3.22}$$

we find

$$A'(t)\cos kx = \left(\frac{1}{2}A^2\cos^2 kx + AB\cos\, kx + \frac{1}{2}B^2\right)_x.$$

Now $\cos^2 kx = (1 + \cos 2kx)/2$, so $(A^2/2)\cos^2 kx = (A^2/4)(1 + \cos\, 2kx)$ and $\left[(A^2/2)\cos^2 kx\right]_x = (-k/2)A^2 \sin 2kx$.

Thus we see that the nonlinearity has generated a second harmonic. For this reason, there can be no solution of the form (3.22) and, by extension, no finite series of the form $\sum_{j=0}^{N} A_j(t)\cos\, jkx$ can be a solution to (3.20) (except the constant solution, i.e., $A_j = 0$ for $j > 0$), and our attempt to find a solution of the form (3.22) must fail. In fact, smooth solutions to (3.20) do not, in general, exist for all time. There are generalized solutions to (3.20) which eventually develop jump discontinuities. These generalized solutions are examples of weak solutions, which we encountered in Section 2.5. These weak solutions are not unique. The Fourier series of a step function has components that are $O(1/k)$ for large k. This guarantees that solutions to (3.20) will have a very rich spectrum.

Analysis of the equation linearized about a constant solution reveals a simple advection process, with speed equal to the (constant) solution. We shall see that a reasonable-looking scheme which is stable according to the linearized analysis is in fact unstable.

Even though there are no nonconstant solutions to (3.20) with finite Fourier series, we can find solutions to the discrete scheme (3.21) of the form

$$u_j^n = c^n \cos \frac{\pi}{2} j + s^n \sin \frac{\pi}{2} j + w^n \cos \pi j + v,$$

where v is a constant and the subscript j and superscript n denote the jth spatial gridpoint and the nth time step.

We then have

$$\begin{aligned} u_{j+1}^n &= -c^n \sin \frac{\pi}{2} j + s^n \cos \frac{\pi}{2} j - w^n \cos \pi j + v, \\ u_{j-1}^n &= c^n \sin \frac{\pi}{2} j - s^n \cos \frac{\pi}{2} j - w^n \cos \pi j + v. \end{aligned}$$

When we substitute this into (3.21) the squared terms cancel. We can calculate the contribution of the cross terms with the aid of the following identities for integer j:

$$\begin{aligned} \sin \frac{\pi}{2} j \times \cos \frac{\pi}{2} j &= 0, \\ \sin \frac{\pi}{2} j \times \cos \pi j &= -\sin \frac{\pi}{2} j, \\ \cos \frac{\pi}{2} j \times \cos \pi j &= \cos \frac{\pi}{2} j, \end{aligned}$$

to arrive at

$$\begin{aligned} c^{n+1} - c^{n-1} &= -4rs^n(w^n - v), \\ s^{n+1} - s^{n-1} &= -4rc^n(w^n + v), \\ w^{n+1} - w^{n-1} &= 0, \end{aligned}$$

where $r = \Delta t / 2\Delta x$.

Now w^n may take on alternate values at alternate steps, say, A and B. Eliminating s yields:

$$c^{n+2} - 2c^n + c^{n-2} = 16r^2(A + v)(B - v)c^n. \tag{3.23}$$

Equation (3.23) is a linear difference equation with constant coefficients, and can be solved by setting c^n equal to a constant p raised to the nth power and solving for p. From this procedure we find that the system is stable, i.e., $|p| \leq 1$, when the coefficient of c^n on the right-hand side of (3.23) is between -4 and 0.

The linearized stability condition derived from the very simplest application of a CFL-like condition to (3.20) would be

$$\frac{\Delta t}{\Delta x} \max |u| < 1.$$

In the leapfrog scheme (3.21), the spatial differences are centered, and thus the spatial differences are taken between points that have indices that are either both even or both odd. For this reason, the term $w^n \cos \pi j$ is effectively a constant for any given value of n in (3.21). For the purposes of stability analysis, one might therefore be tempted to view the terms $c^n \cos(\pi j/2)$ and $s^n \sin(\pi j/2)$ as small deviations from a constant given by $v \pm A$ or $v \pm B$. The strictest CFL condition one could impose would then be $2r[\,|v| + \max\{|A|, |B|\}] < 1$. In this case, the scheme will still blow up if $(A+v)(B-v)$ is *positive*, as we found out from the explicit solution of (3.23), since the equation tends toward the form $c'' - c = 0$.

Schematically this example was intended to mimic a small perturbation about some smooth solution of a larger problem represented by v. The unstable case is precisely the case of significant nonlinearity, i.e., the perturbation u is comparable to v in magnitude.

As noted above, the troubles with this example are usually blamed on aliasing: the $\pi j/2$ terms and the πj term interact to give back something with wavenumber $\approx \pi j/2$. At this resolution, terms proportional to $\sin(3\pi j/2)$ and $\cos(3\pi j/2)$ which would arise from the nonlinear interactions are indistinguishable from $-\sin(\pi j/2)$ and $-\cos(\pi j/2)$ respectively. The most commonly prescribed cure is to use schemes with nice *conservation properties*: schemes in which energy doesn't grow. This will indeed keep the solution from blowing up, but the nature of the problem itself is more subtle than one might gather from a cursory examination.

First of all, there is nothing inherently nonlinear about the instability illustrated above. If we take the difference scheme in question and linearize about $u_j^n = v + w^n \cos \pi j$, we arrive at the same result. Second, if we were to de-alias the scheme perfectly, i.e., if we were to devise some way to eliminate the components with wavenumber $3\pi/2$ that are aliased back to wavenumber $\pi/2$, the result will be a system formally identical to (3.23), with A and B replaced by $A/2$ and $B/2$ respectively, and the instability would remain.

If a conservative method with properties described in the next section, or an artificially dissipative method such as the Lax–Friedrichs or Lax–Wendroff method (also described in the next section) is used, the calculated solution will remain bounded (see Exercise 3.2), but one should be prompted to wonder about the meaning of the numbers so obtained. As noted above, smooth solutions to (3.20) in general cease to exist after some finite time, and the weak solutions, which satisfy (3.20) in some generalized sense, are not unique. The Lax–Friedrichs and Lax–Wendroff methods will produce consistent approximations to solutions of Burgers' equation:

$$u_t - uu_x = \nu u_{xx},$$

for a suitably chosen viscosity ν. It can be shown that in the limit as $\nu \to 0$ the solutions of Burgers' equation approach the weak solution of (3.20), which is compatible with the integral conservation law from which (3.20) can be derived; see, e.g., Whitham (1974). It is easy to show that there are step-like solutions of Burgers' equation, with the jump taking place in a narrow region whose width is inversely proportional to the Reynolds number UL/ν, where U and L are speed and length scales respectively. It is intuitively reasonable that, of the family of generalized solutions to (3.20), one would want to choose the one that is the limiting case of Burgers' equation for small dissipation. The meaning of the (necessarily bounded) solutions of a conservative set of difference equations derived from (3.20) is not clear.

3.4 Conservation laws and conservative difference schemes

Clearly, if a scheme conserves energy, or at least some positive definite functional of the approximate solution, it will not suffer from nonlinear instability. There is also, at first glance, a certain natural appeal to schemes that conserve quantities which are conserved in nature. We begin by examining the nature of conservation laws.

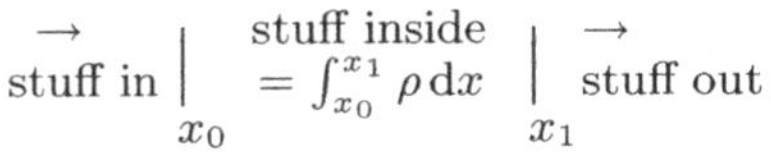

$$\begin{array}{ccccc} \rightarrow & & \text{stuff inside} & & \rightarrow \\ \text{stuff in} & | & = \int_{x_0}^{x_1} \rho\, \mathrm{d}x & | & \text{stuff out} \\ & x_0 & & x_1 & \end{array}$$

Fig. 3.8 Schematic diagram illustrating general one-dimensional conservation of an arbitrary quantity.

We can formulate very general conservation laws. Referring to Figure 3.8, we have

$$\frac{\mathrm{d}}{\mathrm{d}t}(\text{amount of stuff between } x_0 \text{ and } x_1) =$$
$$\text{rate at which stuff comes in} - \text{rate at which stuff goes out.}$$

Let ρ = density of stuff = amount of stuff per unit length. We then have

$$\text{total stuff} = \int_{x_0}^{x_1} \rho \, \mathrm{d}x.$$

Let F(x) = rate at which stuff passes the point x; positive = left to right (units = stuff/second). Then

$$\frac{\mathrm{d}}{\mathrm{d}t} \int_{x_0}^{x_1} \rho \, \mathrm{d}x = F(x_0) - F(x_1).$$

Next divide by $x_1 - x_0$ and let $x_1 - x_0$ go to zero:

$$\frac{\partial \rho}{\partial t} + \frac{\partial F}{\partial x} = 0. \tag{3.24}$$

In higher space dimensions F is a vector quantity:

$$\frac{\partial \rho}{\partial t} + \nabla \cdot \mathbf{F} = 0. \tag{3.25}$$

Our favorite advection equation is a conservation law: write $F = u$:

$$u_t + u_x = 0.$$

The one-dimensional shallow-water equations can also be written in conservation form:

$$\begin{aligned} u_t + \left(\frac{1}{2}u^2 + gh\right)_x &= 0, \\ h_t + (hu)_x &= 0. \end{aligned}$$

It is more physically meaningful to write the shallow-water equations in terms of conservation of momentum and mass:

$$\begin{aligned} (hu)_t + \left[u(hu) + \frac{1}{2}gh^2\right]_x &= 0, \\ h_t + (hu)_x &= 0. \end{aligned}$$

Other useful conservation laws can be derived as shown next.

Begin with the linearized shallow-water equations:

$$\begin{aligned} u_t + gh_x &= 0, \\ h_t + Hu_x &= 0. \end{aligned}$$

Multiply the first by uH and the second by gh:

$$
\begin{aligned}
H\left(\frac{1}{2}u^2\right)_t + ugHh_x &= 0,\\
\left(\frac{1}{2}gh^2\right)_t + gHhu_x &= 0.
\end{aligned}
$$

These two can be combined to form

$$
\left(\frac{1}{2}gh^2 + \frac{1}{2}Hu^2\right)_t + gH(hu)_x = 0, \tag{3.26}
$$

which is the expression for conservation of energy.

We may require that our numerical schemes have desirable conservation properties. We also note that most schemes conserve *something*, as shown by the next example.

Consider, for now, the semidiscrete approximation of $u_t = uu_x$:

$$
\begin{aligned}
\dot{u}_j &= \frac{1}{2\Delta x}\left(\frac{1}{2}u_{j+1}^2 - \frac{1}{2}u_{j-1}^2\right),\\
\sum_{j=1}^{N} \dot{u}_j = \frac{\mathrm{d}}{\mathrm{d}t}\left[\sum_{j=1}^{N} u_j\right] &= \frac{1}{4\Delta x}[u_{N+1}^2 + u_N^2 - u_1^2 - u_0^2],
\end{aligned}
$$

which can be rearranged to form

$$
\Delta x \sum_{j=1}^{N} \dot{u}_j = \frac{1}{2}\left[\frac{1}{2}(u_{N+1}^2 + u_N^2) - \frac{1}{2}(u_1^2 + u_0^2)\right].
$$

As $\Delta x \to 0$, the sum on the left becomes an integral and we recover the conservation equation in integral form. The discretized solution u therefore obeys a conservation law, but this does not imply stability in any reasonable sense, since u is not guaranteed to be a positive quantity. The discrete solution may exhibit large oscillations, while $\sum_{j=1}^{N} u_j$ behaves nicely.

It would be more useful to find an explicit description of the behavior of $\sum_{j=1}^{N} u_j^2$. We may attempt to examine the behavior of this quantity by multiplying the scheme by u_j and summing over the domain. When we do this, we find that

$$
u_j\dot{u}_j = \frac{1}{2}(\dot{u}_j^2) = \frac{u_j}{2\Delta x}\left(\frac{1}{2}u_{j+1}^2 - \frac{1}{2}u_{j-1}^2\right).
$$

The right-hand side is not of the form $F_{j+1} - F_{j-1}$ for any F, and the interior terms will not cancel when the right-hand side is summed over j.

Consider now a staggered grid scheme for the linearized shallow-water equations similar to the one we are using in our reduced gravity model. The equations are

$$\begin{aligned} u_t + gh_x &= 0, \\ h_t + Hu_x &= 0. \end{aligned}$$

The velocity and the height are sampled at alternate points, so $u_{j+1/2}$ is evaluated at $x = (j + 1/2)\Delta x$ and h_j is evaluated at $x = j\Delta x$.

Take the semidiscrete centered scheme:

$$\begin{aligned} \dot{u}_{j+1/2} + \frac{g}{\Delta x}(h_{j+1} - h_j) &= 0, \\ \dot{h}_j + \frac{H}{\Delta x}(u_{j+1/2} - u_{j-1/2}) &= 0. \end{aligned}$$

As in the continuous case, multiply the top equation by $Hu_{j+1/2}$ and the bottom by gh_j and add the two equations to yield

$$\frac{\mathrm{d}}{\mathrm{d}t}\left[\frac{1}{2}Hu_{j+1/2}^2 + \frac{1}{2}gh_j^2\right] + \frac{gH}{\Delta x}\left[u_{j+1/2}h_{j+1} - u_{j-1/2}h_j\right] = 0.$$

Writing $F_j = gHu_{j-1/2}h_j$ the expression becomes

$$\Delta x\frac{\mathrm{d}}{\mathrm{d}t}\left[\frac{1}{2}Hu_{j+1/2}^2 + \frac{1}{2}gh_j^2\right] + F_{j+1} - F_j = 0,$$

and summing from 0 to N yields

$$\Delta x\frac{\mathrm{d}}{\mathrm{d}t}\sum\left[\frac{1}{2}Hu_{j+1}^2 + \frac{1}{2}gh_j^2\right] + F_{N+1} - F_0 = 0.$$

Therefore, the semidiscrete scheme conserves energy.

As $\Delta x \to 0$, the discrete formula converges to

$$\frac{\mathrm{d}}{\mathrm{d}t}\int_{x_1}^{x_1}\frac{1}{2}Hu^2 + \frac{1}{2}gh^2\mathrm{d}x + gHu(x_1,t)\,h(x_1,t) - gHu(x_0,t)\,h(x_0,t) = 0,$$

which is the integral form of the law of conservation of energy.

As this simple example shows, it is possible to devise semidiscrete difference schemes that conserve certain physical invariants exactly. An energy conserving scheme, for example, will not be subject to the instabilities noted in Section 3.3.2 above, though it may not avoid instability in a physically reasonable way. It is not always desirable to sacrifice accuracy or efficiency for conservation properties, but there are some situations in which such schemes are useful. A simple example is a long integration of a model with coarse resolution, in which it might be essential that the numerical simulation not lose a significant amount of mass over the integration time. Several methods with special conservation properties were investigated by Sadourny (1975). This section follows that paper.

Begin by writing the fully nonlinear shallow-water equations:

$$\begin{aligned} u_t + uu_x + vu_y + gh_x &= 0, \\ v_t + uv_x + vv_y + gh_y &= 0, \\ h_t + \nabla \cdot (hu) &= 0. \end{aligned}$$

Sadourny chose $f = 0$ because he performed his integrations on a doubly periodic domain, in which the meaning of the f and β plane approximations is problematical. Here we neglect rotation to follow Sadourny and for simplicity. We begin by deriving the conservation of potential vorticity. The evolution equation for the vorticity $\zeta = v_x - u_y$ is derived by taking the curl of the zonal and meridional momentum equations above:

$$\zeta_t + \mathbf{u} \cdot \nabla\zeta + \zeta\Delta = 0,$$

where $\mathbf{u}$ is the horizontal velocity vector (u, v) as usual and $\Delta = \nabla \cdot \mathbf{u}$. Write

$$\frac{\mathrm{D}}{\mathrm{D}t}\frac{\zeta}{h} = \frac{1}{h^2}\left(h\frac{\mathrm{D}\zeta}{\mathrm{D}t} - \zeta\frac{\mathrm{D}h}{\mathrm{D}t}\right),$$

where $\mathrm{D}/\mathrm{D}t$ denotes the total derivative: $\mathrm{D}/\mathrm{D}t \equiv \partial/\partial t + \mathbf{u} \cdot \nabla$. The right-hand side of this equation is seen to vanish by substitution of the expressions for $\mathrm{D}\zeta/\mathrm{D}t$ and $\mathrm{D}h/\mathrm{D}t$. The conservation of potential vorticity $\eta = \zeta/h$ of each fluid parcel is thus demonstrated.

Multiplication of the evolution equation for η by $h\eta$ and the evolution equation for h by $\eta^2/2$ and adding the two equations yields

$$\left(\frac{\eta^2 h}{2}\right)_t + h\eta\mathbf{u} \cdot \nabla\eta + \left(\frac{\eta^2}{2}\right)\nabla \cdot \mathbf{u}h = 0.$$

We can use the identity $(\eta^2/2)\nabla \cdot \mathbf{u}h = (1/2)\nabla \cdot \eta^2 h\mathbf{u} - \eta h\mathbf{u} \cdot \nabla\eta$ along with Stokes' theorem to derive the conservation equation for *potential enstrophy*:

$$\frac{\mathrm{d}}{\mathrm{d}t}\int_D \left(\frac{\eta^2 h}{2}\right)\mathrm{d}A = -\oint_{\partial D}\left(\frac{\eta^2 h}{2}\right)\mathbf{u} \cdot \mathbf{n}\,\mathrm{d}S,$$

where D is the domain under consideration, ∂D is its boundary and $\mathbf{n}$ is the outer normal to ∂D. The mass (i.e., h) equation is already in conservation form, and conservation of total energy $\int_D (1/2)(h + \mathbf{u} \cdot \mathbf{u})\, h\, \mathrm{d}A$ can be derived directly from the basic equations in a similar fashion, or from the equations in the form used by Sadourny:

$$\begin{aligned}
\mathbf{u}_t + \eta \mathbf{N} \times (h\mathbf{u}) + \nabla \left[h + \frac{1}{2}(\mathbf{u} \cdot \mathbf{u}) \right] &= 0, \\
h_t + \nabla \cdot (h\mathbf{u}) &= 0,
\end{aligned}$$

where $\mathbf{N}$ is the unit vector normal to the plane domain D. These equations are themselves derived from the original shallow-water equations by the vector identity $\mathbf{u} \cdot \nabla \mathbf{u} = (1/2)\nabla(\mathbf{u} \cdot \mathbf{u}) + (\nabla \times \mathbf{u}) \times \mathbf{u}$.

The schemes we will consider are defined on the Arakawa C-grid (cf. Figure 3.5), with the vorticity ζ defined in the center of each grid square, i.e., if the h points have indices (i, j) and the u and v points have indices $(i + 1/2, j)$ and $(i, j + 1/2)$ respectively, then the ζ points will have indices $(i + 1/2, j + 1/2)$. The definitions of these schemes become quite complicated. It is convenient to define differencing and averaging operators as follows:

$$\begin{aligned}
\delta_x q(x, y) &= \frac{1}{2}\left[q\left(x + \frac{1}{2}, y\right) - q\left(x - \frac{1}{2}, y\right) \right], \\
\overline{q}^x(x, y) &= \frac{1}{2}\left[q\left(x + \frac{1}{2}, y\right) + q\left(x - \frac{1}{2}, y\right) \right],
\end{aligned}$$

and similarly for δ_y and $\overline{q}^y$.

A semidiscrete energy conserving scheme can be written as

$$\begin{aligned}
u_t - \overline{\eta \overline{V}^x}^y + \delta_x H &= 0, \\
v_t + \overline{\eta \overline{U}^y}^x + \delta_y H &= 0, \\
h_t + \delta_x U + \delta_y V &= 0,
\end{aligned}$$

where $U = \overline{h}^x u$, $V = \overline{h}^y v$ and $H = h + (\overline{u^2}^x + \overline{v^2}^y)/2$. Potential vorticity η is defined at vorticity points by the quotient of the vorticity and the average of the four surrounding h points. If we define the discrete energy by $E = \sum (h^2 + h\overline{u^2}^x + h\overline{v^2}^y)/2$, we find

$$\mathrm{d}E/\mathrm{dt} = \sum (U u_t + V v_t + H h_t).$$

Substituting from our difference scheme for u_t, v_t and h_t yields

$$\begin{aligned}\mathrm{d}E/\mathrm{d}t &+ \sum(V\eta\overline{\overline{U}^y}^x - U\eta\overline{\overline{V}^x}^y) \\ &+ \sum(U\delta_x H + H\delta_x U) \\ &+ \sum(V\delta_y H + H\delta_y V) = 0.\end{aligned}$$

In a periodic domain, or a domain in which the initial conditions vanish outside of a finite subdomain, each of the three sums in the above expression vanishes separately. In the parenthetical expression in the first of the sums, every product in the left-hand expression eventually appears in the right-hand. For the other two expressions, it is useful to note that one can verify by a simple calculation that there is a product rule for differences which mimics the product rule for derivatives: $\delta_x[UH] = U\delta_x H + H\delta_x U$. This product rule for differences leads to telescoping sums in the remaining two expressions. In a finite domain, there will be non-vanishing boundary terms, corresponding to energy fluxes to or from the exterior of the domain.

The potential enstrophy conserving scheme is similar:

$$\begin{aligned} u_t - \overline{\eta}^y\overline{\overline{V}^x}^y + \delta_x H &= 0, \\ v_t + \overline{\eta}^x\overline{\overline{U}^y}^x + \delta_y H &= 0, \\ h_t + \delta_x U + \delta_y V &= 0. \end{aligned}$$

In the numerical experiments, described by Sadourny, temporal derivatives are discretized by the leapfrog scheme, and all of the dissipation in his simulations comes from relatively infrequent averaging of successive time steps. Dissipation is measured by the frequency at which this averaging process is performed. Some dissipation is necessary to stabilize both the energy conserving and potential enstrophy conserving schemes in their fully discrete forms; rather more dissipation is needed to stabilize the energy conserving scheme.

In the experiments with the energy conserving scheme, both the energy and potential enstrophy start off constant, as they must, and remain so until some critical time. After that time, energy declines steadily and the potential enstrophy rises sharply until it reaches a state of statistical equilibrium, in which it oscillates rapidly about some fixed level which is rather higher than the initial potential enstrophy. This equilibrium value of the enstrophy is fairly insensitive to the value of the dissipation beyond the level required for stabilization. The critical time increases with increasing dissipation. It is clear at this point that the system has reached an artificial balance in which the dissipation is important.

Statistical equilibrium corresponds to equipartition of energy, i.e., the energy is fairly constant as a function of wavenumber, for the energies associated with both the rotational and divergent motion. Increasing resolution increases the critical time and leads to higher equilibrium values of potential enstrophy.

The potential enstrophy conserving scheme behaves rather differently. Both enstrophy and energy decline gradually with time, with the rate of decline decreasing with decreasing dissipation. The equilibrium spectra are also quite different. The rotational energy decreases as k^{-2}, while the divergent energy is relatively constant in wavenumber, above a certain level. The k^{-2} energy spectrum is consistent with a flat potential enstrophy spectrum, since we expect the enstrophy spectrum to go like $k^2E(k)$, where $E(k)$ is the energy spectrum.

Neither of these spectra is terribly realistic. The hope is that we will have enough resolution that the scales of interest are well represented by wavenumbers well below those characterized by equipartition. This tendency toward constant spectra in truncated conservative systems is almost certainly a consequence of truncation. Some motivation for systems such as this one to tend toward constant spectra in the conserved quantity can be found in the book by Chorin (1994), in which it is shown that spectral truncation of the simple conservation equation (3.20) eventually leads to equipartition, and references are given for analogous results for more complex and relevant equations.

Sadourny concludes that the potential enstrophy conserving scheme behaves better than the energy conserving scheme, if for no other reason than its requiring less dissipation to stabilize it, even though the enstrophy conserving scheme will not necessarily produce solutions which remain bounded for all time. It is relevant to point out here that solutions to the nonlinear shallow-water equations may not remain smooth for all time as in the simpler example (3.20).

It is possible to derive schemes that conserve both energy and enstrophy. Such a scheme is described in Chapter 7 of Haltiner and Williams (1980). This is a very complicated scheme, which would be extremely difficult to implement.

Schemes with conservation properties that mirror the physical conservation laws have a great deal of natural appeal, but it is necessary to bear in mind that there are usually schemes which are more accurate for the same level of computing effort. Any convergent scheme will conserve energy and enstrophy up to truncation error. More accurate schemes will yield more accurate phase and group speeds of propagated features. Accurate propagation speeds may be less important than conservation properties for long integrations at coarse resolution, but one must remember that in 1975, when Sadourny's paper was published, the largest supercomputers had less power than even a modest desktop computer available as this is being written. State-of-the-art computing resources allow us to perform long integrations with spatial and temporal resolutions much finer than the finest practical simulations available in the mid-1970s. In today's computing environment, accurate schemes may be a better choice than highly conservative ones.

3.5 Artificial dissipation revisited

We saw in Section 2.4 that the Lax–Friedrichs scheme imposes artificial dissipation at the level of truncation error. We will now derive a related and more accurate scheme, the Lax–Wendroff method, for computation of approximate solutions to conservation laws. As in the Lax–Friedrichs method, artificial dissipation is imposed at the level of truncation error. Of the methods that involve artificial dissipation, it is probably the most widely used.

Consider the conservation equation

$$\mathbf{u}_t + \nabla \cdot \mathbf{F} = 0,$$

where $\mathbf{F}$, the flux, is assumed to be a function of $\mathbf{u}$. The Lax–Wendroff scheme is usually written in two steps: one first takes a (strongly dissipative) Lax–Friedrichs half-step, followed by a leapfrog step. The scheme will be written out for the case of one space dimension here. Thus the first step is

$$\mathbf{u}^*_{j+1/2} = \frac{1}{2}(\mathbf{u}^n_{j+1} + \mathbf{u}^n_j) - \frac{\lambda}{2}[\mathbf{F}(\mathbf{u}^n_{j+1}) - \mathbf{F}(\mathbf{u}^n_j)],$$

and the final step is given by

$$\mathbf{u}^{n+1}_j = \mathbf{u}^n_j - \lambda[\mathbf{F}(\mathbf{u}^*_{j+1/2}) - \mathbf{F}(\mathbf{u}^*_{j-1/2})].$$

For the simple advection equation

$$u_t + u_x = 0,$$

we may combine the two steps of the Lax–Wendroff scheme to form

$$u^{n+1}_j = u^n_j - \frac{\lambda}{2}(u^n_{j+1} - u^n_{j-1}) + \frac{\Delta t^2}{2}\left(\frac{u^n_{j+1} - 2u^n_j + u^n_{j-1}}{\delta x^2}\right).$$

The trailing term is recognizable as a dissipation term. The Lax–Wendroff method is second-order accurate in space and time, and is subject to a CFL condition similar to the leapfrog scheme. Unlike the leapfrog scheme, it does not generate an artificial computational mode.

3.6 Finite-element methods in two space dimensions

The domain can be partitioned into triangles, and local basis functions can be easily defined. Actual construction of these triangulations is a subject in itself, far beyond the scope of this text. Automated triangulation routines are available in packages such as Matlab©. A triangulation of Lake Kinneret, the biblical Sea of Galilee and the only fresh water lake in Israel, is shown in Figure 3.9. Calculation of the requisite inner products of basis functions and their gradients, even for piecewise linear basis functions, can be a daunting task. In practice this is often done by numerical quadratures.

3.7 Open-boundary conditions

In order to understand the proper generalization of the results from Section 2.7, even in one space dimension, we must generalize the notion of characteristics from a single equation to a system. For the purposes of this discussion, it will be useful to write the linearized shallow-water equations in one space dimension as the equations for evolution of small disturbances from a uniform stream with speed U:

$$u_t + Uu_x + gh_x = 0, \tag{3.27}$$

$$h_t + Uh_x + Hu_x = 0, \tag{3.28}$$

which we may write in matrix form as

$$\begin{pmatrix} u \\ h \end{pmatrix}_t + \begin{pmatrix} U & g \\ H & U \end{pmatrix} \begin{pmatrix} u \\ h \end{pmatrix}_x = 0. \tag{3.29}$$

One way to obtain a characteristic form is to multiply (3.29) on the left by the row vector $(l\ m)$. If we choose $(l\ m)$ to be a left eigenvector of the matrix in (3.29), i.e.,

$$(l \quad m) \cdot \begin{pmatrix} U & g \\ H & U \end{pmatrix} = c\,(l \quad m,) \tag{3.30}$$

we find

$$c = U \pm c_0, \tag{3.31}$$

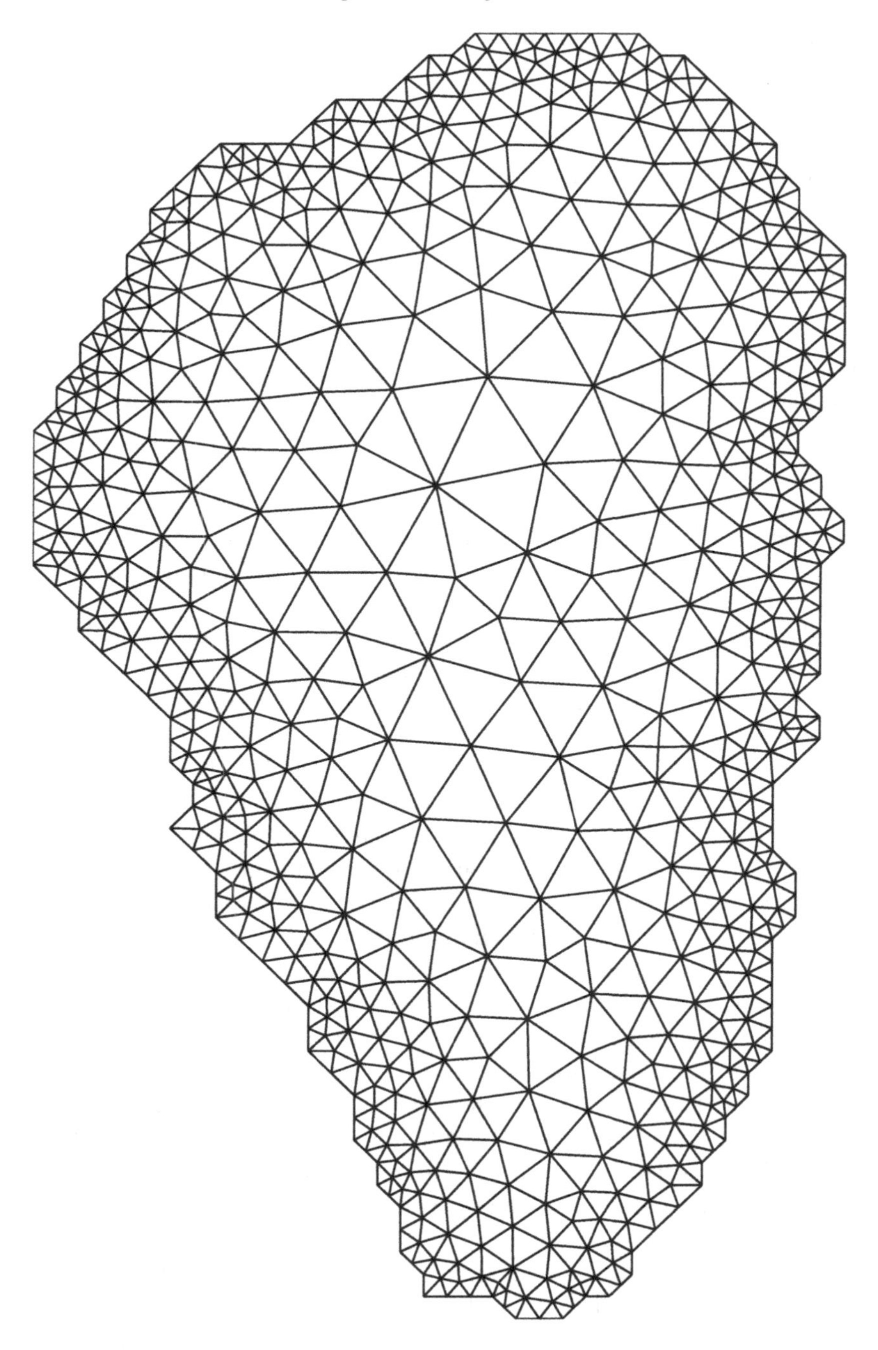

Fig. 3.9 A triangulation of Israel's Lake Kinneret. This triangulation was generated by the partial differential equations package in Matlab©.

where $c_0^2 = gH$. The corresponding eigenvector is given by

$$l = \pm c_0, \tag{3.32}$$

$$m = g. \tag{3.33}$$

Multiplying (3.29) on the left by the row vector $\begin{pmatrix} l & m \end{pmatrix}$ leads to

$$(c_0 u + gh)_t + (U + c_0)(c_0 u + gh)_x = 0 \tag{3.34}$$

for the "+" root and

$$(-c_0 u + gh)_t + (U - c_0)(-c_0 u + gh)_x = 0 \tag{3.35}$$

for the "−" root. Each of these equations is in the same form as the simple advection equation (2.1). We can therefore conclude that the quantity $c_0 u + gh$ is constant along lines in the $x - t$ plane given by $\mathrm{d}x/\mathrm{d}t = U + c_0$ and $-c_0 u + gh$ is constant along lines in the $x - t$ plane given by $\mathrm{d}x/\mathrm{d}t = U - c_0$. The quantities $U \pm c_0$ are called the *characteristic velocities* and $\pm c_0 u + gh$ are known as the *Riemann invariants.* There are a number of equivalent ways to derive this result; details of the approach presented here appear in Whitham (1974).

In order to specify open-boundary conditions for the one-dimensional shallow-water system in an open domain, one must prescribe at the open boundaries a number of independent quantities equal to the number of characteristics that propagate information into the domain. We distinguish two cases for Equation (3.28): if $|U| < c_0$, the flow is said to be *subcritical*; if $|U| > c_0$, the flow is said to be *supercritical.*

In the subcritical case for positive U, at the left-hand boundary, the characteristics with speed $U + c_0$ carry information into the domain, while those with speed $U - c_0$ carry information out of the domain. At the right-hand boundary, the characteristics switch roles. Enough information must therefore be specified on each boundary to determine one of the Riemann invariants, but not both. Specification of both or neither Riemann invariant leads to an ill-posed problem.

In the supercritical case, for positive U, both characteristics carry information in at the left and out at the right; therefore, two independent quantities must be specified at the left-hand boundary, and none must be specified at the right-hand. For negative U, both families of characteristics carry information in at the right-hand boundary and out at the left-hand.

The necessity of correct specification of open-boundary conditions is shown in Figures 3.10–3.14. These figures show the results of computations of initial value problems for the one-dimensional linear shallow-water equations with open boundaries. The resting water depth was given as 10 m, so, with the acceleration of gravity given as $9.87\,\mathrm{m\,s^{-2}}$, the speed c_0 of the waves in the case without advection is about $10\,\mathrm{m\,s^{-1}}$. The domain was 100 m long, divided into 80 intervals. The Lax–Wendroff method with a time step of 0.01 s was used to solve the equation in the interior on an unstaggered grid. Open-boundary conditions were implemented under the assumption that the incoming Riemann invariant vanished. Explicitly setting the incoming Riemann invariants to zero amounts to specifiying one, but not both, of the Riemann invariants at each boundary, as we must in a supercritical flow. This boundary condition is an example of the well-known Sommerfeld radiation condition; see, e.g., Whitham (1974). The outgoing Riemann invariant, i.e., the quantities $c_0 u + gh$ at the right-hand boundary and $-c_0 u + gh$ at the left-hand were calculated by applying upwind schemes to (3.34) and (3.35). With both Riemann invariants thus known at the boundary points, boundary values for u and h can be determined.

Figure 3.10 shows an example of the subcritical case, with the advection speed U set to $5\,\mathrm{m\,s^{-1}}$. At the initial time, h was set to the function shown in the top panel of Figure 3.10, and u was set to zero. The solution is a superposition of two waves, one moving rapidly to the right with speed $U + c_0$, and the other moving more slowly to the left with speed $U - c_0$. The figure clearly shows the disturbance propagating out of the domain as expected, leaving very little behind.

If the same boundary conditions are imposed for the case $U = 11\,\mathrm{m\,s^{-1}}$ the outcome is quite different. In this case the solution is supercritical, and two conditions should be specified at the left-hand boundary and none at the right-hand. Overspecification of the solution at the right-hand boundary leads to rapid oscillations and total amplitudes an order of magnitude greater than the peak amplitude of the true solution, as shown in Figures 3.11 and 3.12. The slower of the two wave modes travels at about $1\,\mathrm{m\,s^{-1}}$, so 60 seconds into the calculation, the slower of the two disturbances should be most of the way out of the domain. Instead we see rapid oscillation with amplitude much larger than is plausible.

If we now implement proper boundary conditions for the supercritical case, we recover a good approximation to the exact solution, as shown in Figures 3.13 and 3.14. This calculation was similar to the earlier one, but both Riemann invariants were set to zero at the left-hand boundary, and both were calculated by upwind schemes at the right-hand. These calculations demonstrate the necessity of imposing proper conditions at open boundaries. The results shown here are evidence that the result

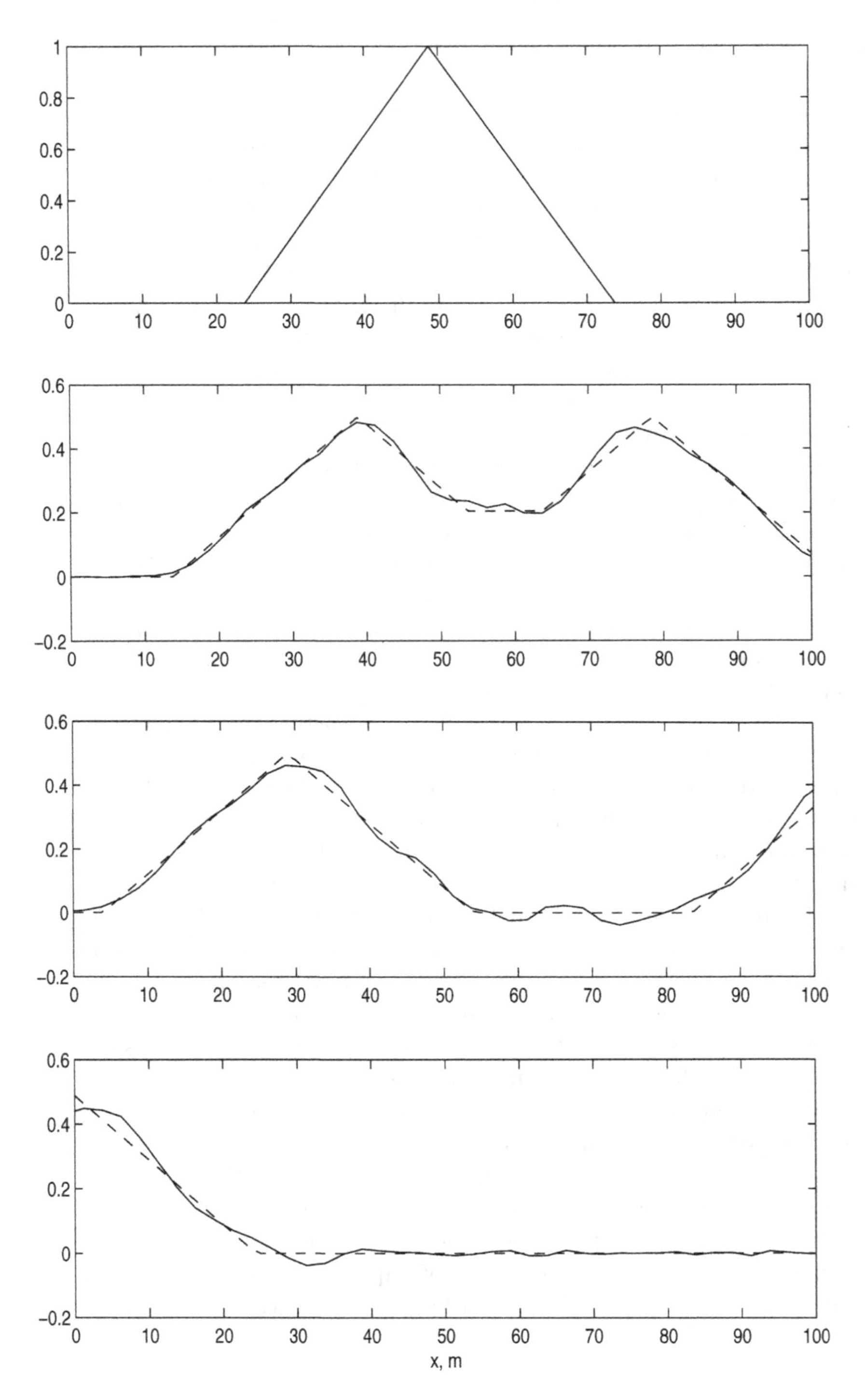

Fig. 3.10 Wave height profiles for solution of the shallow-water equations in the subcritical case. Here, $g = 9.87\,\mathrm{m\,s^{-2}}$, $H = 10\,\mathrm{m}$ and $U = 5$. Computed solutions at a sequence of times are shown as solid lines, while the corresponding exact solutions are shown as dashed lines. The vertical axes are the wave heights in meters. From top to bottom, solutions are shown at times 0 s, 2 s, 4 s and 10 s.

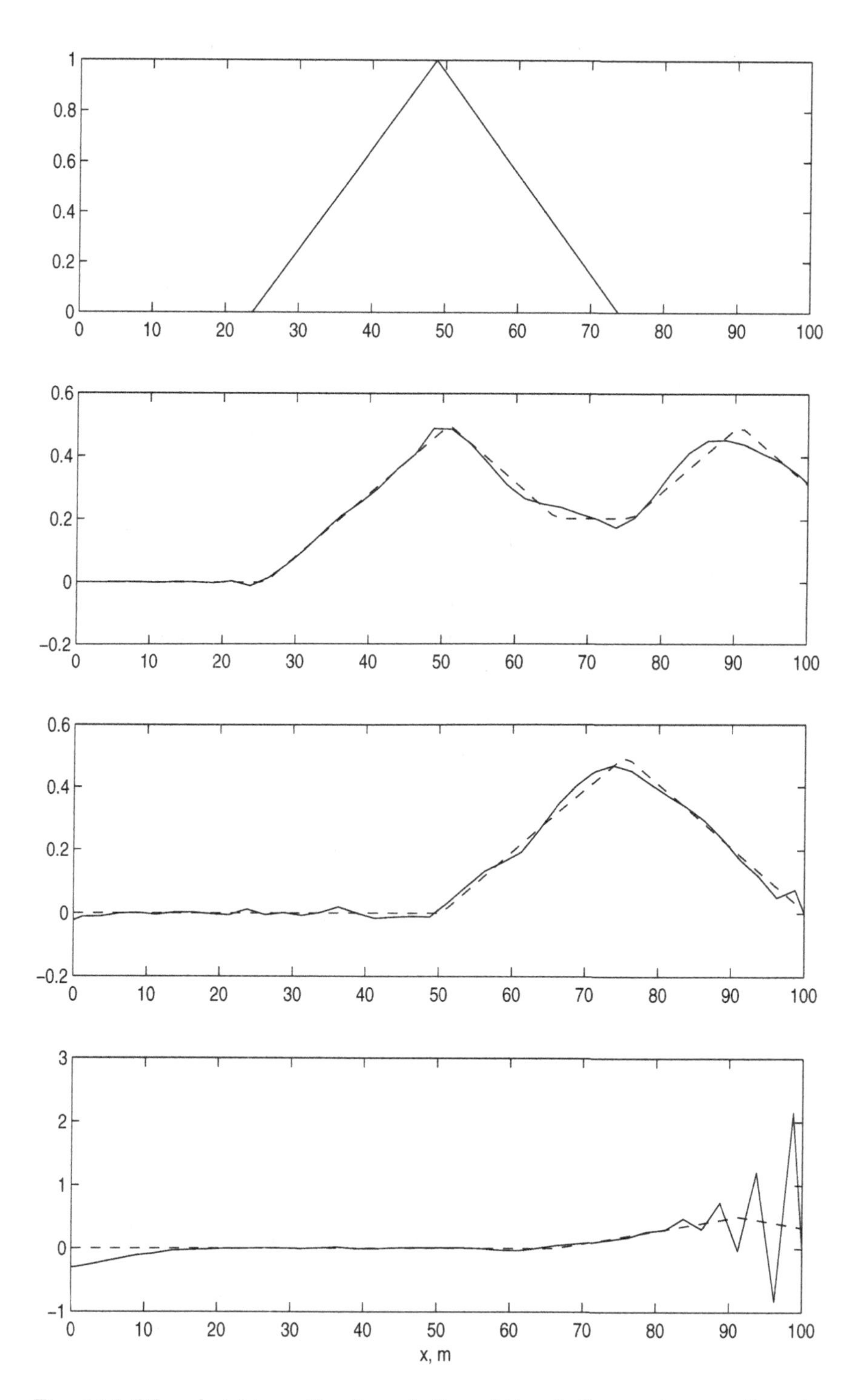

Fig. 3.11 Wave height profiles for solution of the shallow-water equations in the supercritical case with one Riemann invariant prescribed at each boundary. Here, $g = 9.87\,\mathrm{m\,s^{-2}}$, $H = 10\,\mathrm{m}$ and $U = 11$. Computed solutions at a sequence of times are shown as solid lines, while the corresponding exact solutions are shown as dashed lines. The vertical axes are the wave heights in meters. From top to bottom, solutions are shown at times 0 s, 2 s, 25 s and 40 s.

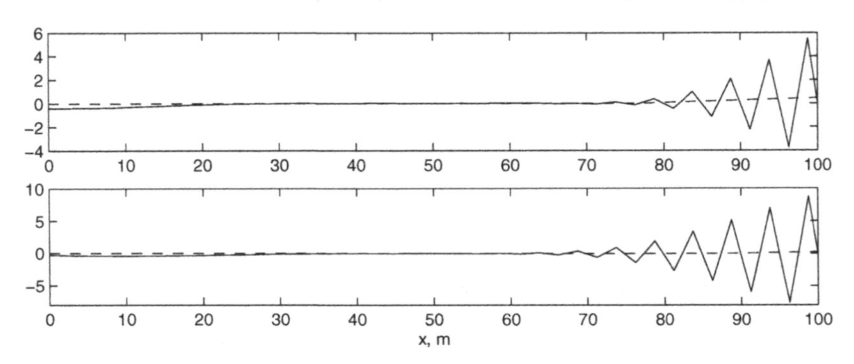

Fig. 3.12 Continuation of Figure 3.11. Top: exact and computed wave heights at 50 s; bottom: exact and computed wave heights at 60 s.

of mis-specification of open-boundary conditions results in an ill-posed computational problem.

The two-dimensional case can be understood in similar terms. We may write the two-dimensional linear shallow-water equations with uniform advection in a form similar to (3.28):

$$u_t + Uu_x + gh_x = 0, \tag{3.36}$$

$$v_t + Uv_x + gh_y = 0, \tag{3.37}$$

$$h_t + Uh_x + Hu_x + Hv_y = 0. \tag{3.38}$$

In order to investigate the well-posedness of the open-boundary problem, it is convenient to derive the expression for conservation of energy. We do this in a manner similar to the derivation of (3.26). Multiply (3.36) by Hu, multiply (3.37) by Hv, multiply (3.38) by gh and add the results to derive

$$E_t + \nabla \cdot (UE + c_0^2 uh, c_0^2 vh) = 0, \tag{3.39}$$

where

$$E = \frac{1}{2}H(u^2 + v^2) + \frac{1}{2}gh^2. \tag{3.40}$$

Now let us consider the problem of (3.38) for positive U in an open channel with rigid walls aligned parallel to the x-axis, so the walls lie along $y = y_0$ and $y = y_1$ between $x = x_0$ and $x = x_1$. Since U is positive, the inflow boundary is $\{(x_0, y) : y_0 < y < y_1\}$ and the outflow boundary is $\{(x_0, y) : y_0 < y < y_1\}$.

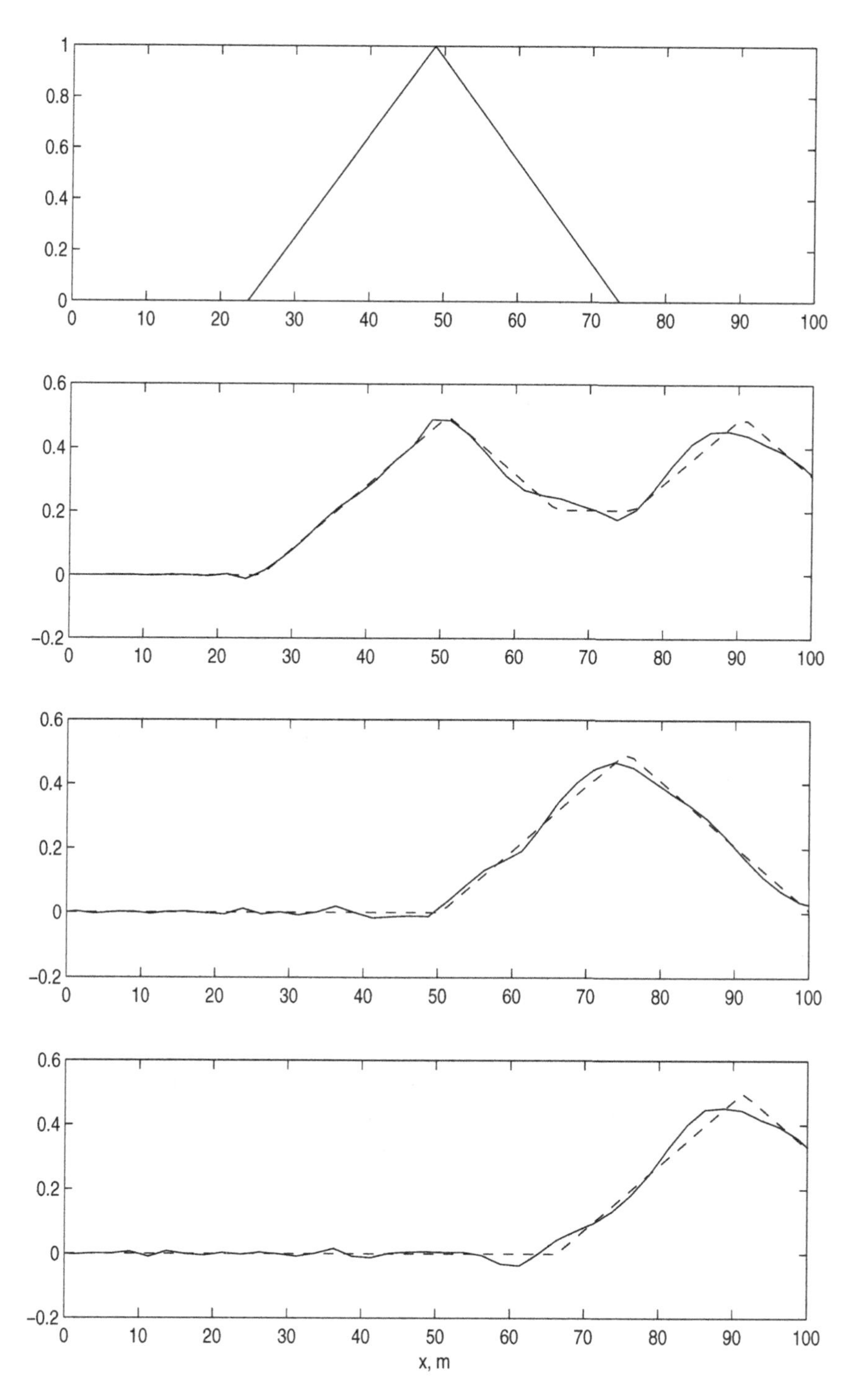

Fig. 3.13 Similar to Figure 3.11, but for the supercritical case with correctly posed boundary conditions.

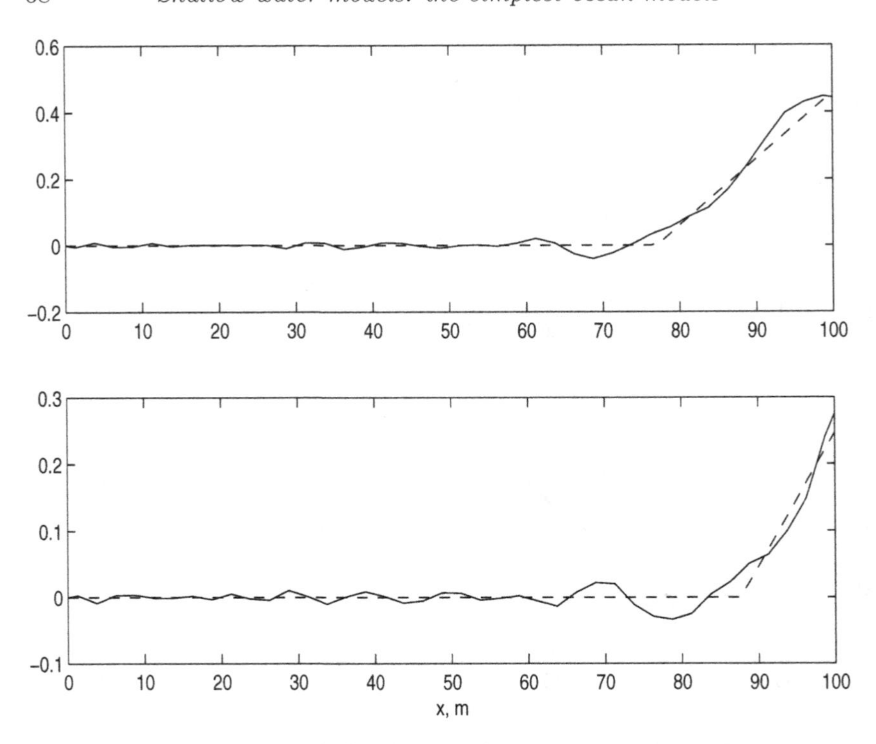

Fig. 3.14 Similar to Figure 3.12, but for the supercritical case with correctly posed boundary conditions.

We can integrate (3.39) over the channel to find

$$\begin{aligned}\frac{\mathrm{d}}{\mathrm{d}t}\int E dA &= -\int \nabla\cdot(UE + c_0^2 uh, c_0^2 vh)\mathrm{d}A\\ &= \oint c_0^2 vh\mathrm{d}x - (UE + c_0^2 uh)\mathrm{d}y,\end{aligned} \tag{3.41}$$

where the line integral in (3.41) is taken over the boundary of the channel. The portion of the integral over the rigid-walls vanishes due to the rigid-wall boundary condition $v(x, y_0, t) = v(x, y_1, t) = 0$ for $\{x_0 \leq x \leq x_1\}$, leaving only the contributions from the inflow and outflow:

$$\frac{\mathrm{d}}{\mathrm{d}t}\int E dA = -\int_{y_0}^{y_1} UE + c_0 uh\,\mathrm{d}y\Big|_{x=x_1} + \int_{y_0}^{y_1} UE + c_0 uh\,\mathrm{d}y\Big|_{x=x_0}. \tag{3.42}$$

We can write the integrands in (3.42) as

$$
\begin{aligned}
UE + c_0uh &= \frac{1}{2}UH(u^2+v^2) + \frac{1}{2}Ugh^2 + c_0uh \\
&= \frac{1}{2}H[u \quad v \quad (c/H)h] \\
&\quad \times \begin{pmatrix} U & 0 & c_0 \\ 0 & U & 0 \\ c_0 & 0 & U \end{pmatrix} \begin{pmatrix} u \\ v \\ (c_0/H)h \end{pmatrix}.
\end{aligned} \tag{3.43}
$$

The eigenvalues of the 3×3 matrix in (3.43) are U, $U + c_0$ and $U - c_0$, with corresponding eigenvectors $\begin{pmatrix}0 & 1 & 0\end{pmatrix}^{\mathrm{T}}$, $\begin{pmatrix}1 & 0 & 1\end{pmatrix}^{\mathrm{T}}$ and $\begin{pmatrix}1 & 0 & -1\end{pmatrix}^{\mathrm{T}}$. If we impose the boundary conditions

$$
\begin{pmatrix} u \\ v \\ (c_0/H)h \end{pmatrix} \cdot \begin{pmatrix} 1 \\ 0 \\ 1 \end{pmatrix} = u + (c_0/H)h = 0 \tag{3.44}
$$

and

$$
\begin{pmatrix} u \\ v \\ (c_0/H)h \end{pmatrix} \cdot \begin{pmatrix} 0 \\ 1 \\ 0 \end{pmatrix} = v = 0 \tag{3.45}
$$

at the inlet $x = x_0$, and

$$
\begin{pmatrix} u \\ v \\ (c_0/H)h \end{pmatrix} \cdot \begin{pmatrix} 1 \\ 0 \\ -1 \end{pmatrix} = u - (c_0/H)h = 0 \tag{3.46}
$$

at the outlet $x = x_1$, we are guaranteed that the right-hand side of (3.42) will be nonpositive, since imposing these conditions assures that the integrand of the first integral be nonnegative and the second be nonpositive. These boundary conditions therefore lead to a well-posed problem. We observe that if $u - (c_0/H)h = 0$ then $c_0(u - (c_0/H)h) = cu - gh = 0$, and if $u + (c_0/H)h = 0$ then $c_0(u + (c_0/H)h) = c_0u + gh = 0$. The reader should recognize $c_0u \pm gh$ as the Riemann invariants for the one-dimensional case (3.28), so the boundary conditions (3.44) and (3.46) mirror the Sommerfeld radiation condition in the one-dimensional case, i.e., incoming waves have zero amplitude, and the last boundary condition, $v = 0$ at $x = x_0$, amounts to a restriction of energetic fluid being advected into the region.

While the shallow-water equations in an open channel with boundary conditions given by (3.44)–(3.46) constitute a well-posed problem that can, in turn, form the basis of a stable computation, solutions to this problem may not be exactly the ones we want. Equations (3.44)–(3.46) are exactly the Sommerfeld radiation conditions for the one-dimensional problem, but the two-dimensional problem admits waves with oblique incidence at the boundary, and these waves may not simply pass out of the domain unimpeded. Consider a wave with wavenumber (k, l), both k and l nonnegative, propagating in the positive direction. Its frequency $\omega = kU + c_0(k^2 + l^2)^{1/2}$ and the solution $(u, v, h)^{\mathrm{T}}$ is proportional to $(kg, lg, \omega - kU)^{\mathrm{T}}$. This only satisfies (3.46) if $l = 0$. Further discussion of this topic can be found in the seminal paper of Engquist and Majda (1977) and in Durran (1999, Chapter 8). The connection between specification of incoming quantities and demonstration of well-posedness by energy calculations, as well as more general criteria for well-posedness is presented in explicit form by Higdon (1986, 1994), including the special case $U = 0$.

3.8 Examples

3.8.1 An early model of basin-scale circulation

One of the earliest models of the development of basin-scale circulation was the shallow-water model presented by Gates (1968). Gates derived his model beginning with the primitive equations

$$u_t + \nabla \cdot (u\mathbf{V}) + (uw)_z + p_x/\rho_0 - fv = A\nabla^2 u + \tau_z^{(x)}/\rho_0, \tag{3.47}$$

$$v_t + \nabla \cdot (v\mathbf{V}) + (vw)_z + p_y/\rho_0 + fu = A\nabla^2 v + \tau_z^{(y)}/\rho_0, \tag{3.48}$$

$$p_z + \rho g = 0, \tag{3.49}$$

where ∇ is the two-dimensional gradient, $\mathbf{V} = (u, v)^{\mathrm{T}}$ is the two-dimensional velocity vector and $\tau^{(x)}$ and $\tau^{(y)}$ are the x- and y-stress components respectively. The free surface is at $z = \zeta(x, y, t)$, at which we impose the free surface boundary condition

$$w\big|_{z=\zeta} = \frac{\mathrm{D}\zeta}{\mathrm{D}t}. \tag{3.50}$$

We integrate Equations (3.47)–(3.49) in depth. We make the approximations

$$\nabla \cdot < u\mathbf{V} > \approx \left(\frac{< u >^2}{\zeta + h}\right)_x + \left(\frac{< u >< v >}{\zeta + h}\right)_x, \tag{3.51}$$

where $< \cdot > \equiv \int_{-h}^{\zeta} \cdot \mathrm{d}z$ ($z = -h$ is the bottom), and neglect bottom stress to find

$$u_t = fv - g(\zeta + h)\zeta_x + A\nabla^2 u + \frac{\tau^{(x)}}{\rho_0} - \frac{(u^2)_x + (uv)_y}{\zeta + h}, \tag{3.52}$$

$$v_t = -fu - g(\zeta + h)\zeta_y + A\nabla^2 v + \frac{\tau^{(y)}}{\rho_0} - \frac{(v^2)_y + (uv)_x}{\zeta + h}, \tag{3.53}$$

$$\zeta_t = -u_x - v_y. \tag{3.54}$$

Here u and v are the integrated velocities, and we must divide by $h + \zeta$ to get velocity components. No-slip conditions are imposed at all boundaries. Boundary values of ζ are calculated by appealing to (3.54) to find $\zeta_t = -(\mathbf{v}_n)_n$, where "$n$" signifies the normal direction, so the right-hand side is the normal derivative of the normal component of the velocity.

The domain is a rectangular basin 5920 km $\times$ 4000 km in zonal and meridional extent respectively. The grid is a staggered grid that does not correspond exactly to any of the grids shown in Figure 3.5. Pressure and velocity stagger in both directions, i.e., there is a velocity point, with both components of horizontal velocity, on either side of every pressure point in both x- and y-directions, i.e., a grid square with pressure points at the corners would have velocity points at the midpoints of each edge. The spacing is 80 km between like points, i.e., 40 km between pressure and velocity points. Time evolution was calculated by the leapfrog scheme. The Coriolis acceleration was given by the usual β-plane approximation, i.e., $f = f_0 + \beta y$, with $f_0 = 0.5 \times 10^{-4}\,\mathrm{s}^{-1}$, the value of the Coriolis acceleration at the southern boundary, given to be at 20° N, and $\beta = 1.75 \times 10^{-11}\,\mathrm{m}^{-1}\,\mathrm{s}^{-1}$, the value at 40° N. The undisturbed depth of the basin was 400 m. The peak wind stress was $0.2\,\mathrm{N\,m}^{-2}$.

Three runs were performed, each with a duration of several months. In all three, dissipation is imposed by a four-point Laplacian, evaluated at the lagged time, as is necessary for stability. In the first, the "basic" experiment, the viscosity coefficient was $A = 10^4\,\mathrm{m}^2\,\mathrm{s}^{-1}$. Two more experiments were performed, both with viscosity coefficient $A = 6 \times 10^3\,\mathrm{m}^2\,\mathrm{s}^{-1}$, one differing from the basic experiment only in the value of A, and the other a fully linear calculation with inertial terms omitted.

The simulations were done with full gravity, so the gravity waves will be faster than the baroclinic waves for an ocean with an active layer of 400 m depth, but slower than the fastest gravity waves in an ocean basin with a depth typical of major ocean basins, say 5000 m or so. With these specifications, the deformation radius will be $\approx 4000^{1/2}\,\mathrm{m\,s^{-1}}/0.5 \times 10^4\,\mathrm{s^{-1}} \approx 1200\,\mathrm{km}$. The Munk layer, with scale $(A/\beta)^{1/3} \approx 83\,\mathrm{km}$ for the basic experiment and 70 km for the experiments with reduced viscosity, is barely resolved. We expect the results to be very diffusive.

Still, the fast waves determine the CFL condition, so $(gh)^{1/2} \approx 63\,\mathrm{m\,s^{-1}}$, and we expect time steps no larger than 1000 s, i.e., 20 minutes or so. The no-slip condition on the boundary is easily implemented. For ζ points on the boundary, we may approximate the normal derivative of the normal velocity by assuming a fictitious velocity point one half-grid interval outside the domain, and assigning a velocity equal in magnitude and opposite in direction to the corresponding point inside the domain. Consider, for example, the pressure gridpoint (x_0, y_k) on the eastern boundary. At time step j, the surface height deflection at (x_0, y_k) is given by ζ_0^k and the depth-integrated zonal velocity at $(x_0 + \Delta x/2, y_k)$ is given by $u_{1/2}^k$. In order that u vanish on the solid boundary, we make the approximation $u_{-1/2}^k = -u_{1/2}^k$, so the centered leapfrog approximation to the continuity equation at the boundary is given by

$$\begin{aligned}\zeta_0^{k+1} &= \zeta_0^{k-1} - \frac{2\Delta t}{\Delta x}(u_{1/2}^k - u_{-1/2}^k)\\ &= \zeta_0^{k-1} - \left(\frac{4\Delta t}{\Delta x}\right) u_{1/2}^k. \end{aligned} \tag{3.55}$$

What do we expect from this simulation? The gravity waves cross the basin in a time given by $4000\,\mathrm{km}/63\,\mathrm{m\,s^{-1}}$, which is approximately 17 hours, so we expect the pressure to set up in a few days. This is borne out by the evolution of the height anomaly in Figure 3.15.

The general pattern of western intensification is set up by day 12. The details of the western boundary current are determined by the interaction of Rossby waves with the western boundary (see, e.g., Pedlosky, 1979), so the structure of the western boundary current will continue to evolve for more than a month, as shown in Figure 3.15.

The development of the circulation can be seen clearly as consisting of wavelike patterns in Figure 3.16 that drift westward. Early in the simulation, the pattern takes on scales typical of the entire basin, and propagates westward with speed roughly that of the Rossby wave with the corresponding wavenumbers. As the simulation progresses, higher wavenumber disturbances appear, indicating the presence of reflected waves with eastward group velocity. The maximum and minimum values

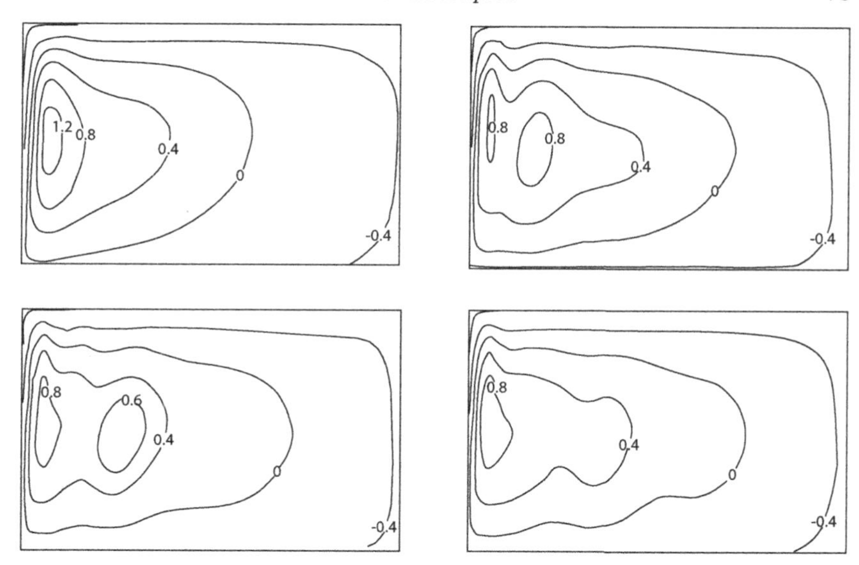

Fig. 3.15 Evolution of height anomaly. Contours are in meters. Top row, left to right: height field after 12 days and 24 days; bottom row, left to right: 36 days and 48 days. Redrawn from Figure 3 of Gates (1968), with permission of the American Meteorological Society.

of v at each wave crest propagate eastward – see the curve drawn in long dashes and dots in Figure 3.17 – at the group velocity predicted by the theory of reflection of Rossby waves from a western boundary (see. e.g., Pedlosky, 1979, Chapter 3).

The motion of the wave crests in the basic experiment is shown in Figure 3.17. In this figure, the maximum meridional velocity appears adjacent to the western boundary early in the experiment, and remains fixed; this is shown by the nearly vertical line at gridpoint number three.

As we might expect, nonlinearity is not terribly important in these simulations. The most evident effect is the southwest–northeast tilt in the wave crests evident in Figure 3.16. In the linear experiment, the wave crests orient nearly exactly north–south, and the north–south asymmetry of the western boundary current region is not so pronounced.

A graph of the total kinetic energy for the basic experiment is shown in Figure 3.18. The total energy rises to a maximum in about eight days, followed by damped oscillations with period typical of the lowest

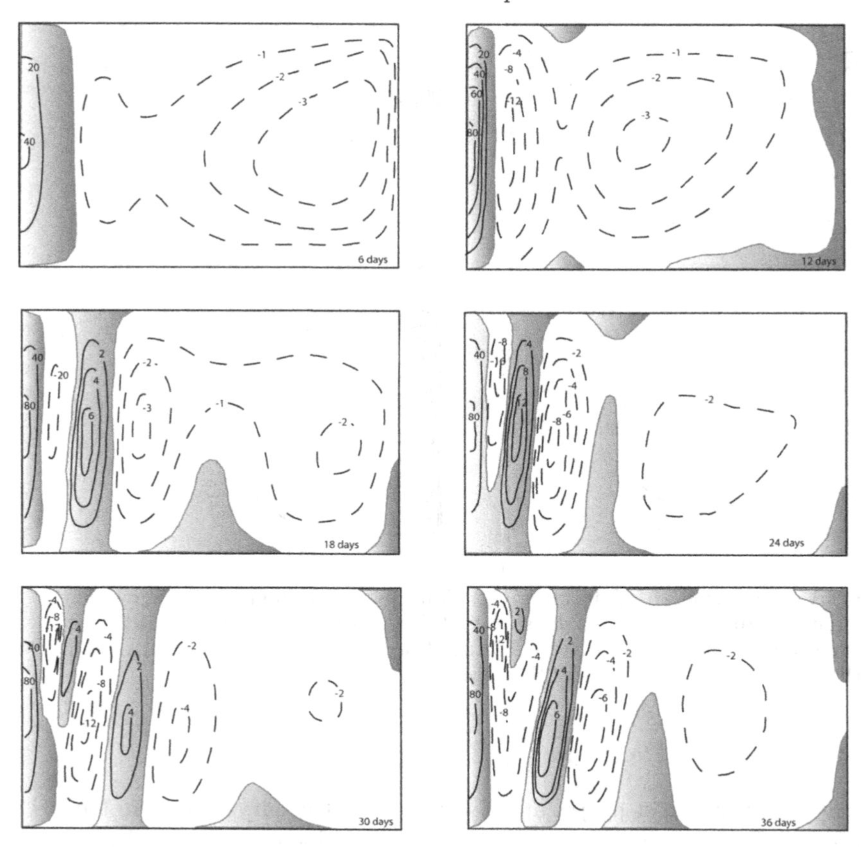

Fig. 3.16 Contour plots of meridional velocity for the basic experiment at 6-day intervals. Contour labels are in units of $10^{-2}\,\mathrm{m\,s^{-2}}$. Regions in which $v \geq 0$ are shaded. Note that contour intervals are not uniform. Redrawn from Figure 4 of Gates (1968), with permission of the American Meteorological Society.

wavenumber quasigeostrophic basin mode (see, e.g., Pedlosky, 1979, Chapter 3). There is little further change in energy after 50 days.

3.8.2 Stable and unstable equilibria in shallow-water models

While simple reduced gravity models are inadequate for simulation of observed ocean circulation, they remain useful as conceptual tools because of the essential features of ocean circulation that they do contain. These models are sufficiently simple that they run economically even

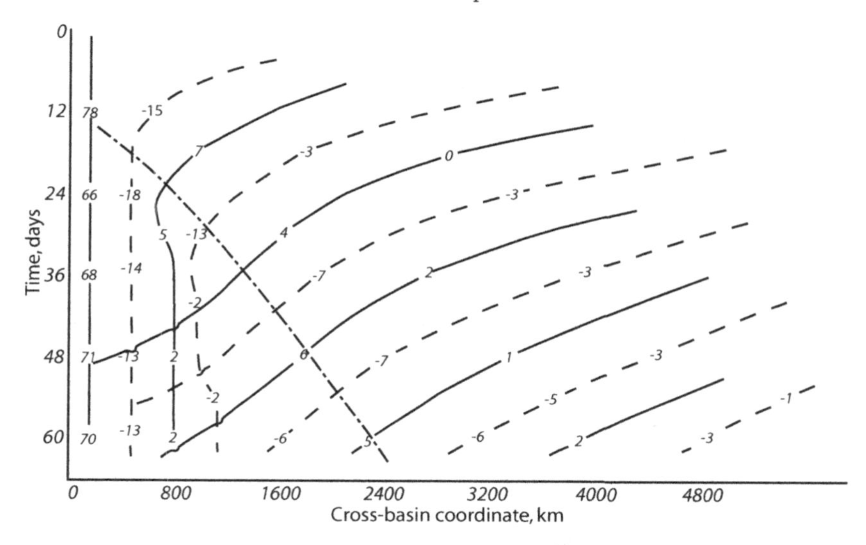

Fig. 3.17 Motion of maxima and minima of meridional velocity in the $x - t$ plane. Solid lines indicate maxima. Dashed lines indicate minima. Numbers are values of the meridional velocity. Long dash–dot curve connects points of maximum velocity on each characteristic. Redrawn from Figure 5 of Gates (1968), with permission of the American Meteorological Society.

on small computers, so long runs with varying parameters may be performed and analyzed in detail. Their relative economy is due in part to the relatively small number of quantities to be calculated, in this case, two velocity components and a layer thickness at each gridpoint, so a simulation with 20 km resolution in a 2000 km × 2000 km basin would involve the calculation of 30 000 quantities at each time step. These 30 000 quantities, i.e., 10 000 copies of each of the two velocity components and the layer thickness, taken together are known as the *state vector*, so such a model would be said to have a *state dimension* of 30 000, as opposed to state dimensions as high as 10^6–10^7 for detailed general circulation models with more complex physics. It is practical, for instance, under some circumstances to calculate the steady states of simple models such as these, and thus understand the qualitative behavior of the system from a dynamical systems viewpoint (see, e.g., Guckenheimer and Holmes, 1983). Finding the roots of a nonlinear system of dimension $O(10^4)$ is hard enough, but finding the roots of a system the size of a general circulation model is beyond the reach of our computational

Table 3.1 Model parameters

Authors	Domain extent	Grid resolution	Time step	Diffusion	Rayleigh friction
Sura *et al.*	2400 km × 2400 km	20 ∗ 91 km	1200 s	$2 \times 10^2\ \mathrm{m^2\,s^{-1}}$	$10^{-7}\ \mathrm{s^{-1}}$
Dijkstra and Molemaker	1000 × 2000	variable FE	–	2×10^2	–
Speich *et al.*	1000 × 2000	20	–	3×10^2	5×10^{-8}
Jiang *et al.*	1000 × 2000	20	1200 s	3×10^2	5×10^{-8}

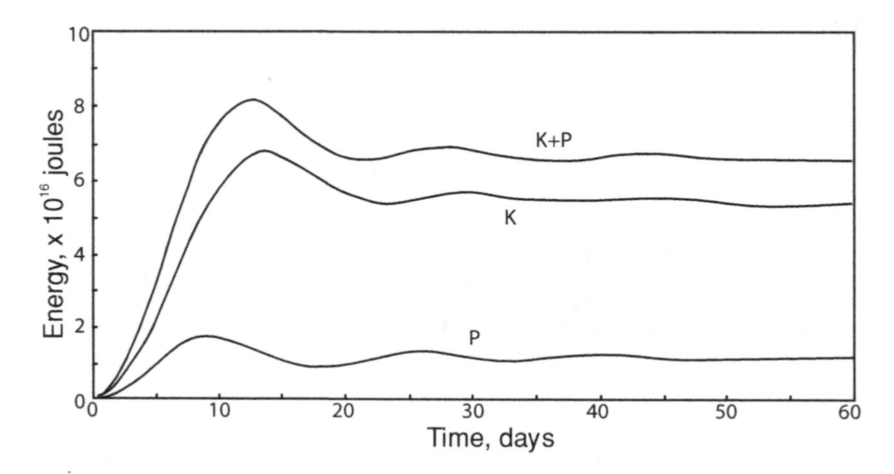

Fig. 3.18 Evolution of kinetic energy, potential energy and total energy. Redrawn from Figure 11 of Gates (1968), with permission of the American Meteorological Society.

facilities, and useful equilibria of general circulation models may not exist anyway.

A sampling of results from these simple models is given here. A review can be found in Dijkstra and Ghil (2005), and references therein. Some authors chose to impose dissipation in the form of Rayleigh friction in addition to diffusion. Rayleigh friction is expressed by terms of the form $-ru, -rv$ on the right-hand side of the x- and y-momentum equations respectively. The parameter r has the dimensions of s^{-1}. Parameters from selected models are given in Table 3.1.

Jiang *et al.* (1995) performed a series of experiments with a reduced gravity model in which they investigated qualitative differences between

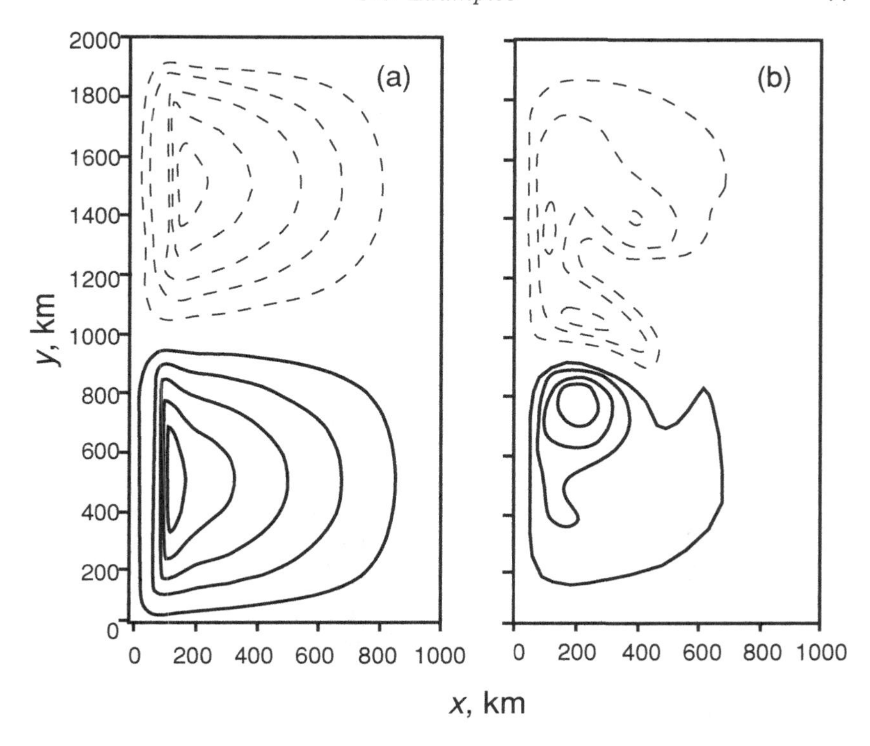

Fig. 3.19 Result of integration of a reduced gravity model driven by steady winds to steady state. Upper layer thicknesses for the linear case (a) and the nonlinear case (b). Redrawn from Figure 1 of Jiang *et al.* (1995), with permission of the American Meteorological Society.

different steady solutions, and documented the existence of periodic solutions, along with the conditions under which they arose. Their model ran in a domain 1000 km in zonal extent and 2000 km in meridional extent. It had a grid resolution of 20 km and a deformation radius of 78 km, and was therefore eddy resolving. Spinning up with fairly weak ($\tau_0 = 0.1\ \mathrm{N\,m^{-2}}$) zonal winds with sinusoidal meridional profile showed significant effect of nonlinearity with the chosen parameters. Figure 3.19 shows the difference between the steady states of the model with and without nonlinearity.

Starting the nonlinear model with different initial conditions and integrating to steady state results in two distinct solutions. The results of the study presented in Jiang *et al.* (1995) are summarized in Figure 3.20.

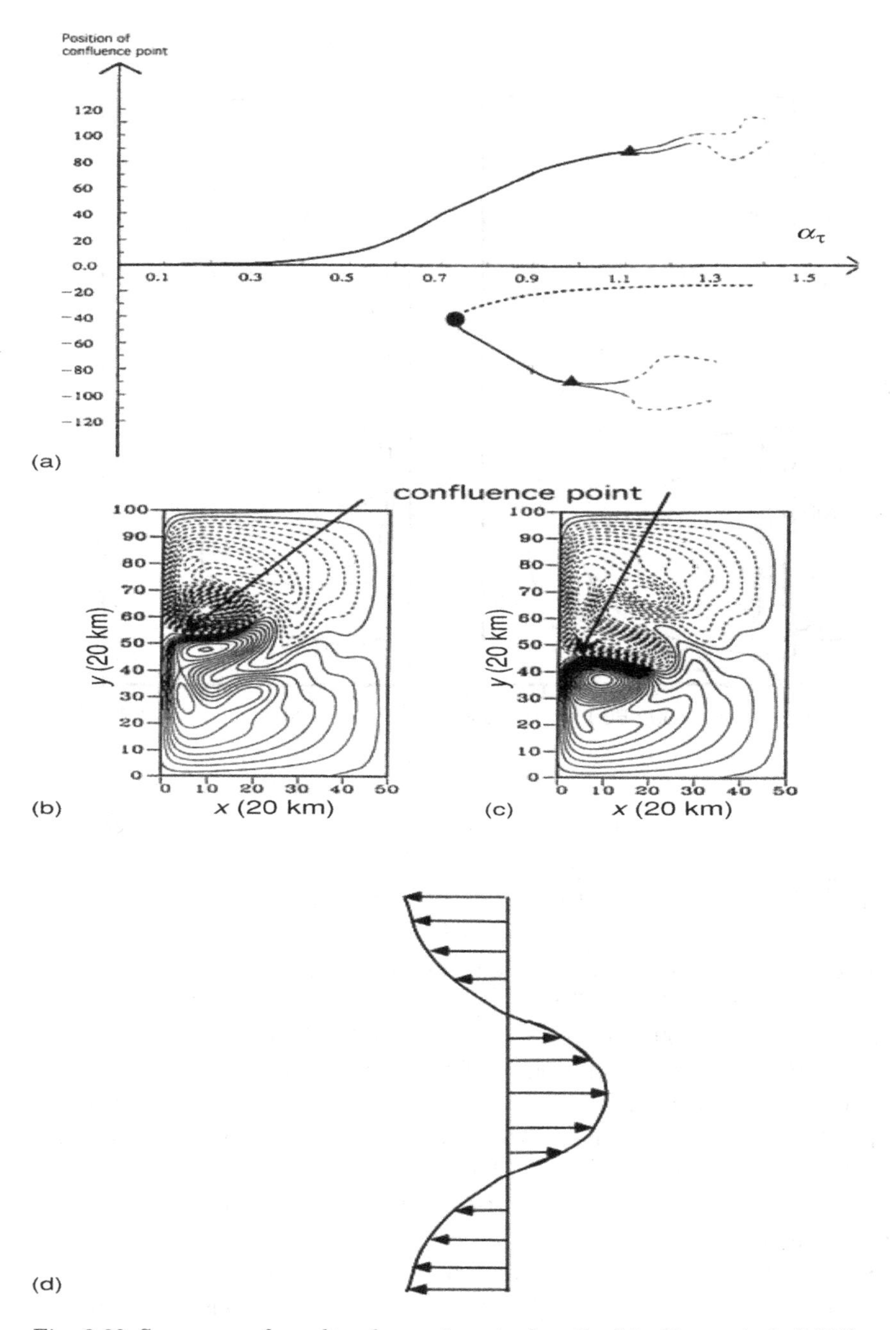

Fig. 3.20 Summary of results of experiments described in Jiang *et al.* (1995) (a). Bifurcation diagram. Each point on the heavy solid curves represents a steady solution to the model equations. The heavy dashed line represents unstable solutions. Light solid and dashed lines represent periodic and aperiodic solutions respectively. The heavy circle and triangles mark bifurcation points. Horizontal axis: wind stress amplitude in $\mathrm{N\,m^{-2}}$; vertical axis: position of confluence of southbound and northbound western boundary currents. (b), (c) Upper layer thicknesses for steady solutions at $\alpha_\tau = 0.9$. Solid contours depict thicknesses greater than 500 m while dashed contours depict thicknesses less than 500 m. (d) Meridional profile of wind stress. Reproduced from Figure 4 of Dijkstra and Ghil (2005), with permission of the American Geophysical Union.

Evidently this system has the property that distinct stable steady states can co-exist; this is clearly impossible in a linear system. These multiple equilibria could only be found when the wind stress amplitude was great enough; solutions corresponding to Figure 3.20(c), with the confluence of the northward and southward western boundary currents occurring south of the center of the basin, could not be found for wind stress amplitudes of less than about $0.7\,\mathrm{N\,m^{-2}}$. In model runs with increased wind stress, Jiang *et al.* found that the stable steady solutions eventually disappeared, to be replaced by periodic solutions. This is evidence of the well known Hopf bifurcation phenomenon (see, e.g., Guckenheimer and Holmes, 1983), i.e., the transition from stability to instability characterized by a conjugate pair of eigenvalues traversing the imaginary axis. In this case, under suitable conditions, a stable periodic solution arises as the steady solution loses stability. As the wind stress increases still further, the periodic solutions are themselves superceded by aperiodic ones.

Speich *et al.* (1995) performed explicit calculations of steady states of the model used by Jiang *et al.* (1995).

Dijkstra and Molemaker (1999) calculated the steady solutions for reduced gravity and quasigeostrophic models in a rectangular basin and in a domain with the geometry of the North Atlantic. Comparison of the results from the two physical approximations will be deferred to Section 5.1. They implemented their reduced gravity model by a finite-element method with triangular elements, with enhanced resolution on the western boundary. They calculated steady states for a range of Ekman numbers from 0.3×10^{-5} to 1.0×10^{-5}, which amounts to varying the momentum transport diffusivity from $150\,\mathrm{m^2\,s^{-1}}$ to $500\,\mathrm{m^2\,s^{-1}}$ while leaving the dimensions latitude of the basin fixed. At the highest Ekman numbers, only a single solution is found, but as the Ekman number decreases an additional pair of solutions, one stable and one unstable, arises, so, for a range of Ekman numbers, two stable solutions co-exist. As the Ekman number decreases, both of these solutions eventually become unstable through Hopf bifurcations. The bifurcation diagrams for the shallow-water model in the "realistic" geometry look similar to those for the rectangular basin, though the solutions in the same range of Ekman numbers for the model in the realistic basin, forced by "realistic" winds (Hellerman and Rosenstein, 1983) are all unstable. Stable branches are restored for the idealized case of sinusoidal winds. This leads the authors to speculate that internal dynamics are more important than geometry in determining the intrinsic variability of the circulation of the ocean.

Sura *et al.* (2001) investigated the effects of stochastic forcing on an idealized double-gyre model. While they did not explicitly calculate steady states of their model, they found that after 10-year spinup with steady winds, the model state settled down to a periodic pattern. Imposition of stochastic forcing led to transition to a qualitatively different regime, which did not persist, and could not be duplicated with unperturbed calculations, but was readily identifiable.

3.9 Choice of computational parameters: an example

An example of model design, from beginning to end, is presented here. As an exercise we follow Bryan (1963); our approach is also similar to Gates (1968). Bryan's goal was to investigate the behavior of an unsteady nonlinear ocean basin model for different choices of parameters; in this case, Rossby number and Reynolds number. We have more computing power and better methods at our disposal than Bryan did, so we should be able to examine regions of parameter space that were inaccessible to Bryan.

Bryan, following Charney *et al.* (1950), who performed the first successful dynamical numerical weather forecast, formulated his model in terms of streamfunction and vorticity, rather than pressure and velocity. Charney *et al.* in their 1950 article explained this choice on the basis of the restrictive CFL condition imposed by gravity waves and the failure of numerical schemes to conserve total vorticity accurately. Here, we will compute pressure and velocity within the framework of a reduced gravity formulation, which, as we have seen, eliminates the surface gravity waves.

In the streamfunction-vorticity formulation, it is not necessary to identify streamfunction directly with pressure, as is done in quasigeostrophic models such as that of Charney *et al.* In Bryan's model, surface gravity waves were eliminated by the use of a rigid lid, i.e., the surface height was held constant, but was allowed to transmit pressure variations. Bryan also assumed a vertical structure in which the velocity was uniform in depth down to a given level, and vanished below that level. This meant that the velocity was nondivergent, and could therefore be derived from a streamfunction. The pressure can be eliminated from the dynamics by taking the curl of the momentum equations (recall that the pressure enters as a gradient, and the curl of a gradient vanishes identically), resulting in a prognostic equation for the vorticity. If desired, the pressure transmitted by the rigid lid can be calculated diagnostically from the velocity field. We shall examine the dynamics of the rigid lid formulation in detail in Chapter 4.

We begin with a linearized reduced gravity model as follows:

$$\begin{aligned} u_t + g'h_x - fv &= \tau(y) + A\nabla^2 u, \\ v_t + g'h_y + fu &= A\nabla^2 v, \\ h_t + H(u_x + v_y) &= 0. \end{aligned} \tag{3.56}$$

We choose parameters roughly corresponding to the North Atlantic:

$$\begin{aligned} H &= 750\,\text{m}, \\ \tau &= -w_0 \cos\left(\frac{\pi y}{2L}\right), \\ L &= 5000\,\text{km}, \end{aligned}$$

$$f = f_0 + \beta y;\ f_0 = 7.3\times 10^{-5}\ \text{s}^{-1};\ \beta = 10^{-11}\ \text{m}^{-1}\ \text{s}^{-1}; g' = 0.02\,\text{m s}^{-2},$$
$$w_0 = 0.2\ \text{N m}^{-2}/10^3\,\text{kg m}^{-3} = 2\times 10^{-4}\,\text{m}^2\,\text{s}^{-2}.$$

We will use a C-grid.

The total region is a rectangle 5000 km in the E–W direction and 10 000 km N–S, as in Bryan (1963). For initial conditions, use Munk's solution with a number of different viscosities, in order to investigate the effect of changes in the viscosity on the behavior of the model. We begin with $A = 5\times 10^3\,\text{m}^2\,\text{s}^{-1}$ as in Stommel (1966, p. 95). We impose no-slip boundary conditions on eastern and western boundaries and free slip on the northern and southern.

What do we expect the answers to look like? There should be no surprises here; the model should behave according to the standard analysis (see, e.g., Stommel, 1966). The interior should behave according to the Sverdrup relation, which is easily derived. The steady motion is given by

$$\begin{aligned} P_x - fv &= \tau^{(x)}, \\ P_y + fu &= \tau^{(y)}. \end{aligned}$$

The pressure may be eliminated to form

$$f(u_x + v_y) + \beta v = \nabla \times \tau,$$

where τ is the vector whose components are $\tau^{(x)}$ and $\tau^{(y)}$. In steady flow, $u_x + v_y = 0$, so $v = \nabla \times \tau/\beta$. In our case,

$$v = -\frac{w_0 \pi}{2\beta L} \sin\left(\frac{\pi y}{2L}\right).$$

From this we have

$$\text{transport} = vLH \approx -\frac{w_0 \pi}{2\beta} H \approx 31 \text{ Sv},$$

where 1 Sverdrup (Sv) $= 10^6\,\text{m}^3\,\text{s}^{-1}$. This is a bit anemic for a Gulf Stream model, but it is about what we expect from such a model as this; again, see Stommel (1966).

The Sverdrup transport is southward in this model. The northward return flow appears in a thin layer along the western boundary. In order to make a proper choice of spatial resolution, we must estimate the scale of the boundary layer.

As noted above, we deduce immediately from the mass equation that the steady flow is nondivergent, and can therefore be represented in terms of a streamfunction ψ with $u = -\psi_y, v = \psi_x$. Substituting this representation of u into (3.9), setting the time derivatives equal to zero and taking the curl of the momentum equations yields

$$A(\psi_{xxxx} + 2\psi_{xxyy} + \psi_{yyyy}) - \beta\psi_x = \frac{\mathrm{d}}{\mathrm{d}y}\tau^{(x)}(y).$$

We could solve this in detail, but here we will simply note that since we are looking for a boundary layer along the western boundary, the zonal scale should be much smaller than the meridional scale. If we scale x by L_x and y by L_y and assume $L_x/L_y << 1$ we obtain

$$A\psi_{xxxx} - \beta\psi_x = O\left(\frac{L_x}{L_y}\right).$$

So we expect scales of the order of $(A/\beta)^{1/3}$. In Munk's (1950) study, he used $A = 5 \times 10^3\ \text{m}^2\,\text{s}^{-1}$; Stommel suggests $10^2\ \text{m}^2\,\text{s}^{-1}$. These yield boundary currents 20–40 km wide; we therefore must resolve scales on that order.

The appropriate scaling for A is the Reynolds number Re, where Re $= UL/A$. U and L are the scale speed and scale length. If we choose U to be $0.01\,\text{m}\,\text{s}^{-1}$ based on the Sverdrup analysis and L to be the basin width of 5×10^6 m, we find

$$A = 5 \times 10^3 \text{ m}^2 \text{ s}^{-1}, \text{ Re} = 10, \text{and}$$
$$A = 10^2 \text{ m}^2 \text{ s}^{-1}, \text{ Re} = 500.$$

The Reynolds number can be viewed as the ratio of the influences of inertial to viscous effects. It is therefore natural to ask about the range of Reynolds numbers in which the linear theory is valid. One may also ask what happens when the linear theory fails. A purely inertial theory (see Stommel, 1966) produces a scale of $(g'H/f^2)^{1/2} \approx 40$ km. This quantity is known as the *radius of deformation*. The question of the range of Reynolds numbers in which the linear theory is valid is an obvious one for consideration with a nonlinear model; clearly a purely linear model cannot answer this question by itself.

In the light of the foregoing discussion, we choose $\Delta x = \Delta y = 25$ km. We use a leapfrog scheme in time for pressure and advection, and we evaluate the diffusion terms at the retarded time step, i.e.,

$$\begin{aligned} u_j^{n+1} = u_j^{n-1} &+ 2\Delta t[\text{discretized form of } - g'h_x^n + fv^n + \tau(y)] \\ &+ 2\Delta t[\text{discretized form of } A\nabla^2 u^{n-1}]. \end{aligned}$$

Were we to use u^n instead of u^{n-1} to calculate the friction term, the result would be an unstable scheme; see Exercise 3.6 or Sod (1985, Chapter 2).

In order to choose a time step, we will examine each term in the equations and investigate the effect of that term alone on the numerical stability of the model.

Our diffusion scheme, taken by itself is

$$u_j^{n+1} = u_j^{n-1} + \frac{2A\Delta t}{\Delta x^2}\,[u_{j+1}^{n-1} - 2u_j^{n-1} + u_{j-1}^{n-1}].$$

We can perform a Fourier analysis of this scheme to determine the maximum time step consistent with stability of the diffusion scheme. This yields

$$\rho(k) = 1 + \frac{2A\Delta t}{\Delta x^2}\,[2(\cos\,k\Delta x - 1)],$$

writing $\lambda = \Delta t/\Delta x^2$,

$$
\begin{aligned}
|\rho(k)| &\leq |1 - 8A\lambda| \\
&\rightarrow 8A\lambda < 2 \\
&\rightarrow \lambda < \frac{2}{8A} \\
&\rightarrow \Delta t < \frac{2\Delta x^2}{8A} = \frac{2 \times (25\,\mathrm{km})^2}{8 \times 5 \times 10^3\,\mathrm{m}^2\,\mathrm{s}^{-1}} \\
&= \frac{2 \times 625 \times 10^6\,\mathrm{m}^2}{40 \times 10^3\,\mathrm{m}^2\,\mathrm{s}^{-1}} \\
&= 31250\,\mathrm{s} \\
&\cong 8.7\,\text{hours}.
\end{aligned}
$$

Next, consider the gravity wave propagation timescales. To do this, examine the pressure terms. Note that there is no advection term in our linear model. Recall the following rough interpretation of the CFL condition: for an explicit method to be stable, the fastest wave must travel no more than Δx in a single time step. The CFL condition for two space dimensions is more stringent than it is in one. The von Neumann analysis for the leapfrog scheme applied to the two-dimensional advection equation $u_t + Uu_x + Vu_y = 0$ leads to a matrix similar in form to (2.12):

$$
\begin{pmatrix} \hat{u}^{n+1} \\ \hat{v}^{n+1} \end{pmatrix} = \begin{pmatrix} -2i\lambda(U\sin k\Delta x + V\sin l\Delta y) & 1 \\ 1 & 0 \end{pmatrix} \begin{pmatrix} \hat{u}^n \\ \hat{v}^n \end{pmatrix} \equiv \hat{G} \begin{pmatrix} \hat{u}^n \\ \hat{v}^n \end{pmatrix}, \tag{3.57}
$$

where we have made the familiar substitution $u = \hat{u}\mathrm{e}^{\mathrm{i}(kx+ly)}$. Analysis similar to that applied to (2.12) leads to

$$
[2(U^2 + V^2)]^{1/2}\lambda < 1 \tag{3.58}
$$

as a condition for stability, so λ is restricted to a value $2^{-1/2}$ times its corresponding value in the one-dimensional case, a reduction of about 30%. This reflects the fact that, for a wave propagating at an angle to the grid, the crest-to-crest distance projected onto the grid axes is greater than it is in the direction of propagation, so the projection of the wave onto the x- or y-axis travels faster than the wave travels in its direction of propagation. For a square grid, i.e., $\Delta x = \Delta y = \Delta$, adjacent crests of a wave with wavelength $2^{1/2}\Delta$, propagating at a 45° angle to the grid (i.e., $k = l$), would appear on the grid 2Δ apart, and thus the apparent phase speed increases by a factor of $2^{1/2}$.

We can make the following rough calculation:

$$c = \sqrt{gH} = \sqrt{0.02\,\mathrm{m\,s^{-2}} \times 750\,\mathrm{m}} \sim 4\,\mathrm{m\,s^{-1}}.$$

Therefore, it takes 6250 s ($\cong$ 1.7 hours) to travel 25 km. We therefore need time steps of 1/2–3/4 hour to be assured of stability. This is the limiting factor in the choice of time steps. With this time step, the inertial oscillations and the dissipation are well resolved.

3.10 Exercises

3.1 The linearized shallow-water equations in one dimension.

(a) Determine the von Neumann condition for the linearized shallow-water equations in one space dimension, using a staggered grid in space and the leapfrog scheme in time.

(b) Use Matlab to implement this method for suitable values of $\lambda = \Delta t/\Delta x$ with periodic boundary conditions on the interval $[0, 2\pi]$. Scale the velocity components by the wave speed c, the height anomaly by the equilibrium layer depth H, length by the deformation radius c/f and time by $1/f$. The result should be a system in which all coefficients are equal to one. Set the initial velocity to zero, and the initial height anomaly to the function used to illustrate the simple advection equation (i.e., the so-called "tent function," or "witch-hat" shown in Figure 2.1). Begin by finding the exact solution to the shallow-water system on a periodic domain with these initial conditions. Run your program for three values of λ, one in the unstable regime and the others in the stable regime. For the values of λ for which the scheme is stable, run your program for a time interval of 2π for two different spatial resolutions. Explain why the results of your five runs look the way they do.

3.2 Consider the Lax–Friedrichs scheme for the simple nonlinear problem given in (3.20):

$$u_j^{n+1} = \frac{1}{2}(u_{j+1}^n + u_{j-1}^n) + \frac{1}{2}\left(\frac{{u_{j+1}^n}^2 - {u_{j-1}^n}^2}{2d}\right).$$

Derive an analog of (3.3.3) for this scheme and comment on its stability properties.

3.3 Implement a linear β-plane reduced gravity model on a 2500 km $\times$ 2500 km square. Use parameters similar to those given in the design example in Section 3.9, but adjusted for the change in basin size. Run your model for 30 days.

(a) Is it in any sort of equilibrium? Plot the kinetic energy/unit volume.

(b) How close is the solution to the traditional Munk/Stommel/Sverdrup picture of the ocean? i.e.,

(i) Where does the Sverdrup balance hold?

(ii) Where is the flow most/least geostrophic?

(c) Decrease the resolution $2\times$. What changes?

(d) Increase the resolution $2\times$. What changes?

(e) Decrease the viscosity by a factor of 10. What changes?

4
Primitive equation models

4.1 Specification of the primitive equation model

The primitive equations, commonly understood as the Navier–Stokes equations in a rotating reference frame, with the vertical momentum equation replaced by the hydrostatic relation, form the basis of the most commonly used models of the ocean on regional to global scales. The flow is generally presumed to be incompressible but stratified; density variations are neglected save in the buoyancy terms. This is known as a shallow Boussinesq approximation. It is much more suitable for the oceans than for the atmosphere.

As in a truly incompressible fluid, there are no exchanges of energy between the flow and the internal energy of any fluid parcel, and a cubic meter of bottom water traveling at $1\,\mathrm{m\,s^{-1}}$ is assumed to have the same momentum as a cubic meter of surface water traveling at the same speed. This has the effect of filtering sound waves, which would otherwise render practical large-scale models impractical.

It is sometimes useful to discard the hydrostatic approximation. This is necessary for the study of the details of deep convection in high latitudes and in some coastal ocean modeling applications. Marshall and co-workers (Marshall *et al.*, 1997a,b) have suggested that nonhydrostatic models may be practical for simulation of the ocean on large scales, given the speed of modern computers. This would have the advantage of a single model for the global circulation, including formation of deep water.

The primitive equations by their nature present a number of obstacles to practical computation. For the most part, practical models differ in the fashion in which they cope with these problems. The most important problems facing all primitive equation models are easily enumerated.

The most natural boundary condition at the surface is the free-surface condition, i.e., the rate of change of the surface elevation is equal to the vertical component of the fluid velocity at the surface. This formulation, while natural, admits the surface gravity waves, which, as we have seen, can impose a severe CFL limitation on explicit calculations. For cases in which the flow in the upper ocean is the focus of interest, one can use the reduced gravity approach introduced in the previous chapter. Other approaches will be described in this chapter.

Because fluid properties are transported more efficiently along isopycnal surfaces than across isopycnal surfaces, diffusion of momentum and of tracers is not isotropic in the ocean on scales of motion commonly treated by ocean models. It is therefore inconvenient to write a diffusion tensor relative to Cartesian coordinates, discretized at depth intervals in the vertical, since isopycnal surfaces are not, in general, exactly horizontal. Models with diagonal diffusion tensors relative to normal spherical coordinates, i.e., latitude, longitude and depth, or approximations to spherical coordinates by Cartesian coordinates in the tangent plane can produce artifacts from excess trans-isopycnal diffusion. Simply assigning different diffusivities in the horizontal and vertical directions is common but unsatisfactory from a rigorous physical viewpoint; see, e.g., Chapter 4 of Pedlosky (1979). The careful reader has already noted that this difficulty is ultimately the result of our lack of understanding of turbulent transport of momentum, heat and chemical tracers. The only real solution to this dilemma is correct representation of turbulent transports. If and when the holy grail of a physically realistic parameterization of the turbulent transport process is found, it will no doubt bring computational difficulties of its own.

One commonly proposed solution to the problem of unrealistic cross-isopycnal diffusion in Cartesian coordinate models is to use the density itself as a vertical coordinate. The result is a so-called layer model, one in which the ocean is modeled as a sequence of slabs in the vertical, each slab being internally homogeneous in all properties; recall the simple one-dimensional example (3.5)–(3.8). In this way, the problems of spurious cross-isopycnal diffusion which are common in Cartesian coordinate models can be avoided, but layer models involve technical problems, as we shall see.

In either case, problems arise in the case of steep topography. Models in which the bottom, or in at least one case, the thermocline, is defined as a coordinate surface have been applied with some success, but have technical problems of their own. Examples of models of this type will be presented in this and subsequent chapters.

Many, if not most, modeling studies are focused on a single region or current system. In those cases, the region of interest may not have natural physical boundaries, and it may be desirable to formulate the problem with open-boundary conditions. Unfortunately, for fundamental physical reasons, most convenient numerical formulations of the open-boundary problem lead to ill-posed problems. Still, some open-boundary conditions have been formulated for practical ocean models that have given rise to stable computations; see Blayo and Debreu (2005) and references therein.

In this chapter the primitive equations and primitive equation models will be developed with the specific intent of performing large-scale simulations. We follow Bryan and Cox (1967) in developing a model suitable for simulating the general circulation of the ocean.

We begin with the 3-D equations of motion; we use β-plane geometry to simplify the notation, but large-scale circulation models actually use spherical geometry. The equations of motion are

$$u_t + uu_x + vu_y + wu_z + \frac{1}{\rho_0} p_x - fv = F^u,$$

$$v_t + uv_x + vv_y + wv_z + \frac{1}{\rho_0} p_y + fu = F^v,$$

$$p(z) = p^s + \int_z^0 g\rho \, \mathrm{d}z',$$

$$p^s = \text{surface pressure},$$

$$u_x + v_y + w_z = 0.$$

The expression for pressure replaces the vertical momentum equation. F^u and F^v represent forcing and/or dissipation.

We have an equation of state for ρ:

$$\rho = \rho(\theta, s, p), \tag{4.1}$$

where θ = potential temperature, s = salinity and p = pressure.

The function on the right-hand side of (4.1) is explicitly specified (Bryan and Cox, 1967). The evolution of θ and s is determined by

$$\theta_t + u\theta_x + v\theta_y + w\theta_z = F^\theta,$$

$$s_t + us_x + vs_y + ws_z = F^s.$$

Turbulent mixing of momentum is often parameterized by eddy viscosity coefficients:

$$F^u = A_{\mathrm{V}} u_{zz} + A_{\mathrm{H}} \nabla^2 u,$$

where ∇^2 is the horizontal Laplacian.

Vertical acceleration is assumed to be negligible in the hydrostatic approximation, so the model cannot deal properly with static instability. Vertical mixing by convection must therefore be parameterized. In the original Bryan–Cox model, turbulent mixing of tracers, including θ, s and passive scalar quantities is parameterized by a convective adjustment:

$$F^T = [(A_{\mathrm{TV}}/\delta) T_z]_z + A_{\mathrm{TH}} \nabla^2 T,$$

where T denotes a general tracer. A_{TV} is the vertical mixing coefficient for T, A_{TH} is the horizontal mixing coefficient. The parameter δ mediates the convective adjustment process, and is given by

$$\begin{pmatrix} \delta = 1 & \rho_z < 0 \\ 0 & \rho_z > 0 \end{pmatrix}.$$

This specifies that the mixing rate is infinite in the statically unstable case. Of course, we cannot actually compute an infinite quantity; the coefficient becoming infinite implies that T_z must be zero, i.e., the tracer concentration immediately becomes vertically uniform. This is basically the assumption that the process of convective adjustment happens instantaneously when viewed on the timescales of interest. Continuous formulations of F^T are available in later versions of the model.

At lateral boundaries, Bryan and Cox impose a no-slip condition. A kinematic boundary condition is imposed at the bottom. If the bottom relief is given by

$$0 = G(x, y, z) = z + H(x, y),$$

then the bottom boundary condition can be written as

$$(u, v, w) \cdot \nabla G = 0,$$

i.e., the velocity at the bottom has no component normal to the bottom. This leads to the condition

$$w + uH_x + vH_y = 0. \tag{4.2}$$

The additional bottom boundary conditions needed for the diffusion operator are given by

$$\rho_0 A_V u_z \Big|_{z=-H} = \tau_B^x,$$

where τ_B^x is the x-component of the bottom stress and A_V is the vertical diffusivity of momentum. A similar equation obtains for y.

Surface stress is imposed at the top:

$$\rho_0 A_V u_z \Big|_{z=0} = \tau^x,$$

If free surface boundary conditions, i.e.,

$$w \Big|_{z=\eta} = \frac{D\eta}{Dt}, \tag{4.3}$$

where $z = \eta(x, y, t)$ is the height of the surface, are imposed, then the resulting formulation admits fast gravity waves, which will, as in the case of shallow-water models examined in the previous chapter, determine the stability conditions for explicit calculations. In order to eliminate the fast gravity waves, the rigid lid condition may be imposed:

$$w \Big|_{z=0} = 0. \tag{4.4}$$

4.2 Dissipation

We model the ocean as a system forced by fluxes of heat and momentum. In order to achieve realistic energetic balances, there must be some way that energy input to the ocean can be dissipated. Mechanical work done on the ocean by the wind must be balanced by some matching frictional process. Seawater is treated as a viscous fluid, but the viscosity of water is, in fact, very small. The kinematic viscosity of water is of the order of $10^{-6}\,\mathrm{m^2\,s^{-1}}$; speeds in western boundary currents are typically $1.0\,\mathrm{m\,s^{-1}}$ and length scales go by the internal radius of deformation, about 40 km in the North Atlantic. The Gulf Stream is therefore characterized by a Reynolds number of 10^{10}, among the highest Reynolds numbers in nature. "Honest" solutions of the Navier–Stokes equations for the ocean, i.e., those derived with the molecular viscosity of water, are beyond contemplation, and even if we could muster the computing resources to perform such calculations, setting them up would raise a host of physical questions about treatment of the boundary, among other things, and we wouldn't know how to interpret the results of such a calculation even if we could do it.

Of course, those scales that contain the apparently random motion that mediates exchanges of momentum, heat and solutes are the realm of turbulent flows. Systematic study of turbulent flow dates to the first half of the twentieth century; entire careers of highly productive scientists have been devoted to it, and it remains an active field of research today. The briefest summary would be far beyond the scope of this book.

Ultimately turbulence theory will give us the key to correct representation of dissipative processes within our models, but at this point we must be satisfied with the admittedly crude representations of these processes. Our attempts to represent dissipative processes in the ocean can only be regarded as desperate.

The simplest, and still widely applied recipe for simulating turbulent effects in models is to assume that the large-scale turbulent transfers resemble the molecular ones, and that one can simply assign an inflated value to the viscosity; this is the so-called "eddy viscosity." We have already encountered artificially large viscosities in Section 3.9.

For large-scale calculations, it is often convenient to assign different eddy viscosity coefficients to the vertical and horizontal directions. Following Pedlosky (1979), one way to do this is to write the stress tensor in the following way:

$$\tau_{xx} = 2\rho A_{\mathrm{H}}\frac{\partial u}{\partial x}; \ \tau_{yy} = 2\rho A_{\mathrm{H}}\frac{\partial v}{\partial y}; \ \tau_{zz} = 2\rho A_{\mathrm{V}}\frac{\partial w}{\partial z}, \tag{4.5}$$

$$\tau_{xy} = \tau_{yx} = \rho A_{\mathrm{H}}\left(\frac{\partial v}{\partial x} + \frac{\partial u}{\partial y}\right), \tag{4.6}$$

$$\tau_{xz} = \tau_{zx} = \rho A_{\mathrm{V}}\frac{\partial u}{\partial z} + \rho A_{\mathrm{H}},\frac{\partial w}{\partial x}, \tag{4.7}$$

$$\tau_{yz} = \tau_{zy} = \rho A_{\mathrm{V}}\frac{\partial v}{\partial z} + \rho A_{\mathrm{H}}\frac{\partial w}{\partial y}. \tag{4.8}$$

This stress tensor is symmetric, as it must be for elementary reasons (see any fluid mechanics text, e.g., Batchelor, 1967), but it has the troubling feature that it associates stress with the rotational component of the rate of strain tensor. Following Batchelor (1967), we decompose the total stress tensor σ into two parts. One is an isotropic part, associated with the pressure p. The other, with which we wish to identify τ above, is referred to as the *deviatoric stress*, i.e., we write $\sigma = pI + \tau$, where I is the 3×3 identity matrix.

The usual assumption about the form of dissipation in a viscous fluid is that the deviatoric stress is related linearly to the rate of strain tensor $\partial u_i/\partial x_j$, where, as usual, $(u_1, u_2, u_3) \equiv (u, v, w)$ and $(x_1, x_2, x_3) \equiv$

(x, y, z). There then must be a fourth-rank tensor A_{ijkl} with

$$\tau_{ij} = \sum_{kl} A_{ijkl} \frac{\partial u_k}{\partial x_l}. \tag{4.9}$$

Now let us examine the form of A_{ijkl} corresponding to (4.5)–(4.8). To simplify the arithmetic, we look at motion confined to the $x - z$ plane. The model (4.5)–(4.8) corresponds to

$$A_{11kl} = \rho \begin{pmatrix} 2A_{\mathrm{H}} & 0 \\ 0 & 0 \end{pmatrix}, \tag{4.10}$$

$$A_{12kl} = A_{21kl} = \rho \begin{pmatrix} 0 & A_{\mathrm{V}} \\ A_{\mathrm{H}} & 0 \end{pmatrix}, \tag{4.11}$$

$$A_{22kl} = \rho \begin{pmatrix} 0 & 0 \\ 0 & 2A_{\mathrm{V}} \end{pmatrix}. \tag{4.12}$$

At this point fluid dynamics texts customarily decompose the rate of strain tensor into a symmetric and an antisymmetric part, i.e.,

$$\frac{\partial u_i}{\partial x_j} = \frac{1}{2}\left(\frac{\partial u_i}{\partial x_j} + \frac{\partial u_j}{\partial x_i}\right) + \frac{1}{2}\left(\frac{\partial u_i}{\partial x_j} - \frac{\partial u_j}{\partial x_i}\right), \tag{4.13}$$

and associate the symmetric part with a pure straining motion and the antisymmetric part with an instantaneous rotation. Isotropy then implies that there is no stress associated with the antisymmetric part, and thus, as an example, no stress associated with rigid body motion, as we expect intuitively. In the case of our model defined by (4.10)–(4.12) we have

$$\sum_{kl} \frac{1}{2} A_{12kl} \left(\frac{\partial u_k}{\partial x_l} - \frac{\partial u_l}{\partial x_k}\right) = (A_{\mathrm{V}} - A_{\mathrm{H}}) \left(\frac{\partial u_1}{\partial x_3} - \frac{\partial u_3}{\partial x_1}\right), \tag{4.14}$$

so there will be stress (but no stress divergence) associated with rigid body motion in the $x - z$ plane given by the familiar form $u = -z$, $w = x$. This reflects the fact that, in choosing different values for vertical and horizontal viscosity, we have given up on the notion of identifying macroscopic scale turbulent transfers with kinematics resembling viscous diffusion of momentum. There is, of course, no reason to expect such an identification to have any rigorous physical justification, but we must be aware that our "eddy viscosities" are not really viscosities at all in the strict sense. We shall see a range of values of A_{H} and A_{V} applied in numerical modeling practice, and all will be orders of magnitude greater than the molecular viscosity of water.

4.3 Dynamics of the rigid lid

As noted in a previous section, application of (4.4) as a boundary condition suppresses the surface gravity waves while admitting pressure variations at the surface. In this section, this approximation is examined in detail.

4.3.1 Derivation of a practical rigid lid PE model

The total pressure p is given by the sum of the surface pressure and the hydrostatic pressure:

$$p = p^s + \int_z^0 g\rho \, \mathrm{d}z, \tag{4.15}$$

write

$$u_t = u_t' - \frac{p_x^s}{\rho_0}, \tag{4.16}$$

$$v_t = v_t' - \frac{p_y^s}{\rho_0}, \tag{4.17}$$

so

$$u_t' = -(\mathbf{u} \cdot \nabla)u + fv - \frac{g}{\rho_0} \int_z^0 \rho_x \, \mathrm{d}z + F^u, \tag{4.18}$$

$$v_t' = -(\mathbf{u} \cdot \nabla)v - fu - \frac{g}{\rho_0} \int_z^0 \rho_y \, \mathrm{d}z + F^v, \tag{4.19}$$

now write

$$u = \hat{u} + \bar{u}, \;\; v = \hat{v} + \bar{v},$$

where the overbar denotes vertical average, e.g.,

$$\bar{u} = \frac{1}{H} \int_{-H}^0 u \, \mathrm{d}z.$$

Then

$$\hat{u}_t = u_t - \bar{u}_t = u_t' - \bar{u}_t', \tag{4.20}$$

since the surface pressure term cancels in the expression $u_t - \bar{u}_t$. A similar equation holds for v. Note that the surface pressure is not a function of z, so $\bar{p}^s = p^s$.

Integrating the equation of continuity implies that

$$0 = \int_{-H}^{0} u_x + v_y + w_z \, \mathrm{d}z$$
$$= \int_{-H}^{0} u_x \, \mathrm{d}z + \int_{-H}^{0} v_y \, \mathrm{d}z + w(0) - w(-H).$$

Now

$$\int_{-H}^{0} u_x \, \mathrm{d}z = \frac{\partial}{\partial x} \int_{-H}^{0} u \, \mathrm{d}z - u(-H)H_x,$$

and

$$\int_{-H}^{0} v_y \, \mathrm{d}z = \frac{\partial}{\partial y} \int_{-H}^{0} v \, \mathrm{d}z - v(-H)H_y.$$

Imposing the boundary condition (4.2) along with the rigid lid assumption $w(0) = 0$ results in

$$\frac{\partial}{\partial x} \int_{-H}^{0} u \, \mathrm{d}z + \frac{\partial}{\partial y} \int_{-H}^{0} v \, \mathrm{d}z = 0.$$

So

$$(H\bar{u})_x + (H\bar{v})_y = 0.$$

In nearly all cases of interest we can define a transport streamfunction ψ with

$$H\bar{u} = -\psi_y; \qquad H\bar{v} = \psi_x.$$

In regions with islands, the transport streamfunction will not be unique. Now

$$(\bar{v}_x - \bar{u}_y)_t = (\bar{v}'_x - \bar{u}'_y)_t,$$

since all pressure terms are eliminated by the cross differentiation, so

$$\left[\left(\frac{\psi_x}{H} \right)_x + \left(\frac{\psi_y}{H} \right)_y \right]_t = (\bar{v}'_x - \bar{u}'_y)_t. \tag{4.21}$$

The boundary condition for ψ is that $\psi =$ constant on lateral boundaries.

In a closed basin with no islands, we may simply take $\psi = 0$ on the boundary. If islands are present, things become rather more complicated; ψ must be constant along the island boundary, but we must allow for the possibility that the value of ψ on the island boundary may not be zero. Equation (4.21) is a Poisson type equation for ψ_t and therefore can be used to predict ψ in terms of the evolution of u' and v'.

This closes the problem. We begin by using (4.18) and (4.19) to calculate u'_t and v'_t. We can then average u' and v' in the vertical and use the results to calculate $\hat{u}_t$ from (4.20) and a similar equation for $\hat{v}_t$. We can then calculate ψ from (4.21) and combine the result with $\hat{u}$ and $\hat{v}$ to get u and v. p^s could be backed out of u' and u by application of (4.16) and (4.17).

4.3.2 Examples with simple dynamics

Consider (our old friend) the linearized reduced gravity system:

$$\begin{aligned} u_t + g'h_x - fv &= 0, \\ v_t + g'h_y + fu &= 0, \\ h_t + H(u_x + v_y) &= 0. \end{aligned}$$

Seek solutions of the form

$$u = u_0 \mathrm{e}^{\mathrm{i}(kx+ly-\omega t)}; \; v = v_0 \mathrm{e}^{\mathrm{i}(kx+ly-\omega t)}; \; h = h_0 \mathrm{e}^{\mathrm{i}(kx+ly-\omega t)}.$$

As before, we find out that $\omega = 0$ or $\omega = \pm\sqrt{gH(k^2+l^2)+f^2}$.

Rigid lid:

$$\begin{aligned} u_t + P_x - fv &= 0, \\ v_t + P_y + fu &= 0, \\ u_x + v_y &= 0. \end{aligned}$$

Take the curl of the momentum equations:

$$(v_x - u_y)_t + f(u_x + v_y) = (v_x - u_y)_t = 0. \tag{4.22}$$

Since $u_x + v_y = 0$, we may define a streamfunction with

$$u = -\psi_y; \quad v = \psi_x.$$

Equation (4.22) then becomes

$$(\nabla^2\psi)_t = 0.$$

If we look for a wavelike solution for ψ: $\psi \propto \mathrm{e}^{\mathrm{i}(kx+ly-\omega t)}$ we find

$$\mathrm{i}\omega(k^2 + l^2)\psi = 0, \tag{4.23}$$

so for real k and l, we must have $\omega = 0$ for nontrivial solutions. We then have, from the equations, $\psi = P/f$. The rigid lid approximation has suppressed the gravity waves, leaving only the steady geostrophically balanced solution. One might be tempted to seek trapped-wave solutions with real l and imaginary k, but one quickly realizes that solutions of the form $\psi = \exp[i(ly - \omega t)]\sinh(x)$ cannot be reasonably normalized.

Let us now look at a slightly more complicated example. Write the shallow water equations:

$$\begin{aligned}
u_t + uu_x + vu_y + \Psi_x - fv &= 0,\\
v_t + uv_x + vv_y + \Psi_y + fu &= 0,\\
\Psi_t + (u\Psi)_x + (v\Psi)_y &= 0,
\end{aligned}$$

where we have written $gh = \Psi$, the geopotential. Now write

$$\begin{aligned}
u &\leftarrow U + u,\\
v &\leftarrow v,\\
\Psi &\leftarrow \Psi_0 - fUy + \varphi.
\end{aligned}$$

Assume $fUy/\Psi_0 = O(u, v, \varphi)$ and that u, v, φ are functions of x and t. The linearized system is

$$\begin{aligned}
u_t + Uu_x + \varphi_x - fv &= 0,\\
v_t + Uv_x + fu &= 0,\\
\varphi_t + U\varphi_x + \Psi_0 u_x - fUv &= 0.
\end{aligned}$$

This system supports three families of waves. The dispersion relation is given by:

$$k^2(U - c)\{(\Psi_0 + f^2/k^2) - (U - c)^2\} - f^2U = 0 \tag{4.24}$$

where $c = \omega/k$, the phase speed. This cubic equation yields three values for c, given approximately (see Exercise 4.2) by

$$c_\pm = U \pm \sqrt{\Psi_0 + f^2/k^2} + \frac{f^2U}{2(k^2\Psi_0 + f^2)}, \tag{4.25}$$

$$\text{and } c_R = U - \frac{f^2U}{k^2\Psi_0 + f^2}. \tag{4.26}$$

The $c_\pm$ are the inertia-gravity waves; these are basically our old friends the shallow water waves. The third, the much slower one, is the Rossby wave. For the Rossby wave, the relationship among the components u, v and φ is given approximately by

$$\begin{pmatrix} u \\ v \\ \varphi \end{pmatrix} = \begin{pmatrix} 1 \\ \frac{\mathrm{i}k\Psi_0}{fU}(1 + \frac{f^2}{k^2\Psi_0}) \\ \frac{\Psi_0}{U} + \frac{f^2(\Psi_0 - U^2)}{k^2 U \Psi_0} \end{pmatrix}. \tag{4.27}$$

Now consider the rigid lid:

$$u_t + Uu_x + P_x - fv = 0,$$
$$v_t + Uv_x + fu = 0,$$
$$\Psi_0 u_x - fUv = 0,$$

so $v = (\Psi_0/fU)u_x$, and the second equation becomes

$$\left(\frac{\Psi_0}{fU}\right) u_{xt} + \left(\frac{\Psi_0}{f}\right) u_{xx} + fu = 0,$$

or

$$u_{xt} + Uu_{xx} + \left(\frac{f^2 U}{\Psi_0}\right) u = 0.$$

Write $u = \mathrm{e}^{\mathrm{i}(kx-\omega t)}$; we then find that

$$\omega k - k^2 U + \frac{f^2 U}{\Psi_0} = 0,$$

hence

$$c_p = \omega/k = U - \frac{f^2 U}{k^2 \Psi_0}.$$

To complete the problem, we calculate the relationship among u, v and P. We begin by arbitrarily setting $u = 1$. Then

$$v = \left(\frac{\Psi_0}{fU}\right) u_x = \mathrm{i}k \left(\frac{\Psi_0}{fU}\right) \mathrm{e}^{\mathrm{i}(kx-\omega t)}. \tag{4.28}$$

P can be derived by substituting into the zonal momentum equation:

$$(-\mathrm{i}\omega + \mathrm{i}Uk) - \mathrm{i}k\frac{\Psi_0}{U} + P_x = 0$$

$$P_x = \left[\omega - Uk + \frac{k\Psi_0}{U}\right] \mathrm{e}^{\mathrm{i}(kx-\omega t)}$$

$$\rightarrow P = \left[c_p - U + \frac{\Psi_0}{U}\right] \mathrm{e}^{\mathrm{i}(kx-\omega t)}.$$

Note that the rigid lid Rossby wave speed differs from the simple case with no rigid lid. This effect will be small if $f^2/k^2\Psi_0 << 1$. Since Ψ_0 is the square of the speed c of the shallow water waves, this says that the rigid lid approximation is reasonable if the spatial scales of interest are much smaller than $\Psi_0^{1/2}/f$, the *external* deformation radius. It is difficult to derive the rigid lid approximation from scaling considerations alone.

Comparison of the expression for v in (4.28) with v in (4.27) shows that the rigid lid cross-stream velocity v differs by a factor of $1+f^2/k^2\Psi_0$ from its free surface counterpart. It is especially important to note that the rigid lid pressure P and the geopotential ϕ from (4.27) also agree to $O(f^2/k^2\Psi_0)$. This illustrates the fact that the pressure P in the rigid lid approximation corresponds to the surface variability in the free-surface formulation. While the surface is nominally flat in rigid lid models, the pressure P is the appropriate quantity for comparison to sea surface height data such as that derived from satellite altimetry.

So the rigid lid has filtered the fast gravity waves, but it has distorted the speed of the Rossby waves. Some discussion of the consequences of this change in the Rossby wave speed may be found in Gates (1968).

4.3.3 Stability restrictions

One potential problem with the rigid lid approximation arises through a combination of steep topography and unequal grid spacing in the x and y directions. This can appear naturally in models of the ocean at high latitudes, where an equally spaced grid in spherical coordinates will have finer grid spacing in the zonal than in the meridional direction. Interestingly, the problem involves the diffusion term in the evolution equation for the transport streamfunction. The following example is adapted from the example presented by Killworth (1987). As in Killworth's example, we neglect advection and consider only the contribution of diffusion to the time derivative on the right-hand side of (4.21). We begin with the semidiscrete approximation, in which the partial differential equation is discretized in y and t, but not in x. We consider a simple step topography, independent of x, with $H = H_2$ for $y \geq y_J$, and $H = H_1$ for $y < y_J$. Our grid is staggered in y, as a one-dimensional version of the B-grid used in the Bryan–Cox model. The geometry of this example is shown in Figure 4.1.

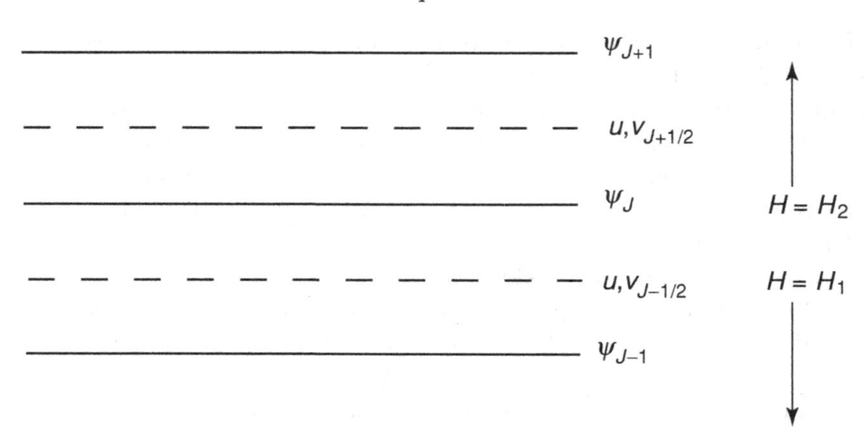

Fig. 4.1 Staggered grid in y for rigid lid example. Velocity components u and v are evaluated at points with half-integer indices. The streamfunction ψ is evaluated at points with integer indices. The depth of the water changes abruptly from H_1 for $y \le y_{J-1/2}$ to H_2 for $y \ge y_J$.

We write (4.21) in this case as

$$\left[\left(\frac{1}{\overline{H}^y}\right)\delta_t\psi_x\right]_x + \delta_y\left[\left(\frac{1}{H}\right)\delta_y\delta_t\psi\right] = (\overline{G^v}^y)_x - \delta_y G^u), \tag{4.29}$$

$$G^u = A_{\rm H}(\partial_x^2 + \delta_y\delta_y)u;\ G^v = A_{\rm H}(\partial_x^2 + \delta_y\delta_y)v. \tag{4.30}$$

The transport streamfunction ψ has dimensions $\mathrm{m^3\,s^{-1}}$. $A_{\rm H}$ is the eddy viscosity, with dimensions $\mathrm{m^2\,s^{-1}}$. In this example, we choose $\psi = 0$ for $y \ne y_J$ and we assume a solution of sinusoidal form in the x direction and exponential growth in time, i.e., after N time steps,

$$\psi_j = \phi_j \mathrm{e}^{\mathrm{i}kx}\zeta^N;\ \phi_j = \Phi\delta_{jJ}, \tag{4.31}$$

for some scalar ζ to be determined. Here δ_{jJ} is the Kronecker δ. The difference operators $\delta_{x,y}$ and the averaging operators denoted by the overbar are as they were defined in Section 3.4. The constant Φ has the dimensions of $\mathrm{m^3\,s^{-1}}$. By our definition of streamfunction, we have

$$u_{J+1/2} = \frac{\Phi e^{ikx}\zeta^N}{H_2\Delta y}; \; u_{J-1/2} = -\frac{\Phi e^{ikx}\zeta^N}{H_1\Delta y}, \tag{4.32}$$

$$v_{J+1/2} = \frac{ik\Phi e^{ikx}\zeta^N}{2H_2}; \; v_{J-1/2} = \frac{ik\Phi e^{ikx}\zeta^N}{2H_1}, \tag{4.33}$$

$$\begin{aligned} G^u_{J+1/2} &= A_{\mathrm{H}}\left(-k^2 u_{J+1/2} + \frac{-2u_{J+1/2} + u_{J-1/2}}{\Delta y^2}\right) \\ &= A_{\mathrm{H}}\left(\frac{-k^2}{H_2\Delta y} - \frac{2}{H_2\Delta y^3} - \frac{1}{H_1\Delta y^3}\right)\Phi e^{ikx}\zeta^N, \end{aligned} \tag{4.34}$$

$$\begin{aligned} G^u_{J-1/2} &= A_{\mathrm{H}}\left(\frac{-k^2(-1)}{H_1\Delta y} + \frac{1}{H_2\Delta y^3} - \frac{2(-1)}{H_1\Delta y^3}\right)\Phi e^{ikx}\zeta^N \\ &= A_{\mathrm{H}}\left(\frac{k^2}{H_1\Delta y} + \frac{1}{H_2\Delta y^3} + \frac{2}{H_1\Delta y^3}\right)\Phi e^{ikx}\zeta^N, \end{aligned} \tag{4.35}$$

$$G^v_{J+1/2} = A_{\mathrm{H}}\left(\frac{-ik^3}{2H_2} + \frac{-ik}{H_2\Delta y^2} + \frac{ik}{2H_1\Delta y^2}\right)\Phi e^{ikx}\zeta^N, \tag{4.36}$$

$$G^v_{J-1/2} = A_{\mathrm{H}}\left(\frac{-ik^3}{2H_1} + \frac{ik}{2H_2\Delta y^2} - \frac{ik}{H_1\Delta y^2}\right)\Phi e^{ikx}\zeta^N, \tag{4.37}$$

$$\begin{aligned} \overline{G^v}^y &= \frac{A_{\mathrm{H}}}{2}\left[\frac{-ik^3}{2}\left(\frac{1}{H_2} + \frac{1}{H_1}\right) + \frac{ik}{2\Delta y^2}\left(\frac{-1}{H_2} - \frac{1}{H_1}\right)\right]\Phi e^{ikx}\zeta^N \\ &= \frac{-ikA_{\mathrm{H}}}{4}\left(\frac{1}{H_2} + \frac{1}{H_1}\right)\left(k^2 + \frac{1}{\Delta y^2}\right)\Phi e^{ikx}\zeta^N, \end{aligned} \tag{4.38}$$

$$\begin{aligned} \delta_y G^u &= A_{\mathrm{H}}\left[\frac{-k^2}{\Delta y^2}\left(\frac{1}{H_2} + \frac{1}{H_1}\right) - \frac{3}{\Delta^4}\left(\frac{1}{H_1} + \frac{1}{H_1}\right)\right]\Phi e^{ikx}\zeta^N \\ &= \frac{-A_{\mathrm{H}}}{\Delta y^2}\left(\frac{1}{H_2} + \frac{1}{H_1}\right)\left(k^2 + \frac{3}{\Delta y^2}\right)\Phi e^{ikx}\zeta^N, \end{aligned} \tag{4.39}$$

$$\overline{G^v}^y_x = \frac{A_{\mathrm{H}}k^2}{4}\left(\frac{1}{H_2} + \frac{1}{H_1}\right)\left(k^2 + \frac{1}{\Delta y^2}\right)\Phi e^{ikx}\zeta^N, \tag{4.40}$$

$$\begin{aligned} \overline{G^v}^y_x - \delta_y G^u &= A_{\mathrm{H}}\left(\frac{1}{H_2} + \frac{1}{H_1}\right)\left(\frac{k^4}{4} + \frac{k^2}{4\Delta y^2} + \frac{k^2}{\Delta y^2} + \frac{3}{\Delta y^4}\right)\Phi e^{ikx}\zeta^N \\ &= A_{\mathrm{H}}\left(\frac{1}{H_2} + \frac{1}{H_1}\right)\left(\frac{k^4}{4} + \frac{5k^2}{4\Delta y^2} + \frac{3}{\Delta y^4}\right)\Phi e^{ikx}\zeta^N. \end{aligned} \tag{4.41}$$

The terms in the above equations have the dimensions s^{-2}, as they should. With the assumption that the solution has the form given in (4.31), the left-hand side of (4.29) becomes

$$\frac{2}{H_1+H_2}(-k^2\Phi \mathrm{e}^{\mathrm{i}kx})\delta_t\zeta^N + \frac{\delta_t}{\Delta_y}\left(-\frac{\Phi \mathrm{e}^{\mathrm{i}kx}\zeta^N}{H_2\Delta y} - \frac{\Phi \mathrm{e}^{\mathrm{i}kx}\zeta^N}{H_1\Delta y}\right)$$
$$= \left[\frac{-2k^2}{H_1+H_2} - \frac{1}{\Delta y^2}\left(\frac{1}{H_2}+\frac{1}{H_1}\right)\right]\Phi \mathrm{e}^{\mathrm{i}kx}\zeta^N \tag{4.42}$$

where $\delta_t\zeta^N \equiv (\zeta^{N+1}-\zeta^N)/\Delta t$. So

$$-\left[\frac{2k^2}{H_1+H_2} + \frac{1}{\Delta y^2}\left(\frac{1}{H_2}+\frac{1}{H_1}\right)\right](\zeta^{N+1}-\zeta^N) \tag{4.43}$$
$$= A_{\mathrm{H}}\Delta t\left(\frac{1}{H_2}+\frac{1}{H_1}\right)\left(\frac{k^4}{4}+\frac{5k^2}{4\Delta y^2}+\frac{3}{\Delta y^4}\right)\zeta^N. \tag{4.44}$$

Hence

$$-\left(\frac{2k^2H_1H_2}{(H_1+H_2)^2}+\frac{1}{\Delta y^2}\right)(\zeta^{N+1}-\zeta^N)$$
$$= A_{\mathrm{H}}\Delta t\left(\frac{k^4}{4}+\frac{5k^2}{4\Delta y^2}+\frac{3}{\Delta y^4}\right)\zeta^N. \tag{4.45}$$

Finally,

$$\zeta^{N+1} = \zeta^N\left[1 - \frac{A_{\mathrm{H}}\Delta t\left(\frac{k^4}{4}+\frac{5k^2}{4\Delta y^2}+\frac{3}{\Delta y^4}\right)}{\left(\frac{2k^2H_1H_2}{(H_1+H_2)^2}+\frac{1}{\Delta y^2}\right)}\right]. \tag{4.46}$$

After all that algebra, we can identify a few special cases:

1. $k=0$, i.e., ψ constant in the x direction: (4.46) becomes

$$\zeta^{N+1} = \zeta^N\left(1 - \frac{A_{\mathrm{H}}\times 3\Delta t}{\Delta y^2}\right). \tag{4.47}$$

The condition for strict stability is then

$$\frac{A_{\mathrm{H}}\Delta t}{\Delta y^2} \le \frac{2}{3}. \tag{4.48}$$

This resembles the usual condition for stability of the initial value problem for the heat equation (see Exercise 2.6).

2. $H_1 = H_2 = H$: The expression in the denominator of (4.46) becomes

$$\frac{k^2}{2}+\frac{1}{\Delta y^2}.$$

Write $k = s/\Delta x$, for fixed interval Δx and dimensionless factor s. By the form of (4.31), we need not consider values of s greater than π. Equation (4.46) then becomes

$$\begin{aligned}\zeta^{N+1} &= \zeta^N \left[1 - \frac{A_{\mathrm{H}}\Delta t\left(\frac{s^4}{4\Delta x^4} + \frac{5s^2}{4\Delta x^2\Delta y^2} + \frac{3}{\Delta y^4}\right)}{\left(\frac{s^2}{2\Delta x^2} + \frac{1}{\Delta y^2}\right)}\right] \\ &= \zeta^N \left[1 - \frac{A_{\mathrm{H}}\Delta t}{\Delta x^2}\frac{\left(\frac{s^4}{4} + \frac{5s^2}{4}\frac{\Delta x^2}{\Delta y^2} + 3\frac{\Delta x^4}{\Delta y^4}\right)}{\frac{s^2}{2} + \Delta x^2 \Delta y^2}\right]. \end{aligned} \tag{4.49}$$

For $s = 1$ and $\Delta x = \Delta y$ we recover the previous case.

3. Finally, suppose $H_2 >> H_1$, $k = s/\Delta x$. Equation (4.46) then becomes

$$\zeta^{N+1} = \zeta^N \left[1 - \frac{A_{\mathrm{H}}\Delta t}{\Delta x^2}\frac{\frac{s^4}{4} + \frac{5s^2}{4}\frac{\Delta x^2}{\Delta y^2} + 3\frac{\Delta x^4}{\Delta y^4}}{2s^2\frac{H_1}{H_2} + \frac{\Delta x^2}{\Delta y^2}}\right]. \tag{4.50}$$

The minimal condition for stability is then

$$\frac{A_{\mathrm{H}}\Delta t}{\Delta x^2}\frac{\frac{s^4}{4} + \frac{5s^2}{4}\frac{\Delta x^2}{\Delta y^2} + 3\frac{\Delta x^4}{\Delta y^4}}{2s^2\frac{H_1}{H_2} + \frac{\Delta x^2}{\Delta y^2}} \leq 2, \tag{4.51}$$

i.e.,

$$\frac{A_{\mathrm{H}}\Delta t}{\Delta x^2} \leq 2\frac{2s^2\frac{H_1}{H_2} + \frac{\Delta x^2}{\Delta y^2}}{\frac{s^4}{4} + \frac{5s^2}{4}\frac{\Delta x^2}{\Delta y^2} + 3\frac{\Delta x^4}{\Delta y^4}}. \tag{4.52}$$

From this it is clear that the stability of the scheme is controlled by the larger of H_1/H_2 and $\Delta x/\Delta y$. Passing from the semidiscrete to the fully discrete case can be accomplished by noting that as differentiating the right-hand side of (4.31) is equivalent to multiplication by $\mathrm{i}k$, calculating the corresponding centered difference quotient $\delta_x \exp(\mathrm{i}kx)$ is equivalent to multiplying by $\mathrm{i}\sin(k\Delta x)/\Delta x$, so the largest value s can assume is 1.0. In the case of an evenly spaced grid, the condition for stability will not be highly restrictive, even in the case of very steep topography. Killworth (1987) noted that a problem arose at high latitudes. In a model formulated in spherical coordinates, for an evenly spaced grid in latitude and longitude, $\Delta x/\Delta y$ will be the cosine of the latitude. Hence Killworth was able to induce a transition from stability to instability in a run of the Bryan–Cox model simply by moving the southern boundary of an Antarctic simulation southward. Clearly the same end can be accomplished by smoothing the topography, so H_1/H_2 is closer to unity. A similar condition might be encountered in certain situations in the coastal ocean.

This is the source of the often-repeated suggestion that the rigid lid model contains an instability in the case of steep topography. Killworth (1987) also notes that fast barotropic waves in the presence of steep topography could impose a more stringent CFL condition than first internal mode gravity waves, but the author is unaware of any systematic study of this phenomenon. For now, we note that the well understood stability restriction on rigid lid models arises as a combination of influences of unequal grid spacing and steep topography. For this reason, many large-scale ocean simulations are run with a free-surface boundary condition. These models deal with the inevitable CFL restrictions by using split-step methods, or treating the free surface implicitly (e.g., Dukowicz and Smith, 1994).

4.4 Spinup of 3D models

In this section we will study the dynamical consequences of stratification in a simplified model. We refer to Anderson *et al.* (1979) for a description of spinup of the full PE model. We begin with the linearized equations of motion:

$$u_t + p_x - fv = 0, \tag{4.53}$$

$$v_t + p_y + fu = 0, \tag{4.54}$$

$$u_x + v_y + w_z = 0, \tag{4.55}$$

$$\rho = \bar{\rho}[\rho_0(z) + \rho_1(x, y, z, t)], \tag{4.56}$$

$$\rho_t + w\rho_0' = 0, \tag{4.57}$$

where $\bar{\rho}$ is the overall mean density. The pressure in (4.53)–(4.54) is given by (4.15), where the surface pressure p^s will be determined later. We have seen one case of a system with vertical structure, i.e., the two-layer model introduced in Section 3.1.

One way to examine the effect of vertical structure is to seek solutions of the equations by the classical method of separation of variables, which is described below.

In order to get it to work, we choose

$$w = W(x, y, t)F(z), \tag{4.58}$$

$$u = U(x, y, t)F'(z), \tag{4.59}$$

$$v = V(x, y, t)F'(z), \tag{4.60}$$

$$\rho_1 = R(x, y, t)F''(z). \tag{4.61}$$

The equations then become

$$\begin{aligned} U_t - gR_x - fV &= 0, \\ V_t - gR_y + fU &= 0, \\ U_x + V_y + W &= 0, \\ R_t F'' + \rho_0' W F &= 0, \end{aligned} \tag{4.62}$$

and evaluation of the integral in (4.15) leads to

$$\nabla[p^s + RF'(0)] = 0. \tag{4.63}$$

Take the horizontal divergence of the original equations:

$$(u_x + v_y)_t + (p_{xx} + p_{yy}) - f(v_x - u_y) + \beta u = 0,$$

add w_{zt} to both sides:

$$p_{xx} + p_{yy} - f(v_x - u_y) + \beta u = w_{zt},$$

write in separated form:

$$-gR_{xx} - gR_{yy} - f(V_x - U_y) + \beta U = W_t,$$

use (4.62) to eliminate W_t:

$$-gR_{xx} - gR_{yy} - f(V_x - U_y) + \beta U = -\frac{R_{tt}F''}{\rho_0' F},$$

so

$$\frac{gR_{xx} + gR_{yy} + f(V_x - U_y) - \beta U}{R_{tt}} = -\frac{F''}{\rho_0' F}.$$

Thus

$$F'' - \lambda(-\rho_0')F = 0. \tag{4.64}$$

For the present purpose, let us assume that the bottom is flat, so the bottom boundary condition is

$$F(-H) = 0. \tag{4.65}$$

For the rigid lid, the upper boundary condition is

$$F(0) = 0. \tag{4.66}$$

Write $N^2 = -g\rho_0'$; we then have $-\rho_0' = N^2/g$ and (4.61) becomes

$$F'' - \lambda \frac{N^2}{g} F = 0.$$

We must have $\lambda < 0$ (or we couldn't arrange the boundary conditions), and λ/g must have the dimensions of $\mathrm{m}^{-2}\,\mathrm{s}^{-2}$; therefore write $-\lambda/g = c^{-2}$. The equation then becomes

$$F'' + \frac{N^2}{c^2}F = 0. \tag{4.67}$$

We can now write a single evolution equation for the amplitude of a single vertical mode. Apply (4.67) to (4.62) to form a single expression proportional to $F(z)$, and use the continuity equation to eliminate W, to obtain

$$U_t - gR_x - fV = 0, \tag{4.68}$$

$$V_t - gR_y + fU = 0, \tag{4.69}$$

$$R_t - (c^2/g)(U_x + V_y) = 0. \tag{4.70}$$

This is formally identical to the shallow-water system with wave speed c. The solutions to the linearized primitive equations can thus be characterized as a superposition of vertical dynamical modes. The vertical structure of each mode is the solution to the eigenvalue problem (4.64). The amplitude of each mode is the solution to the shallow-water equations, with speed given by the eigenvalue in (4.64). The surface pressure p^s is calculated from (4.63).

If N were constant, F would have the form

$$F = \sin\left(\frac{j\pi z}{H}\right)$$

to fit the boundary conditions, so

$$\frac{j^2\pi^2}{H^2} = \frac{N^2}{c^2} \text{ for integer } j \geq 0,$$

and so

$$c^2 = \frac{N^2H^2}{j^2\pi^2}.$$

Recall that $N^2 = -g\rho_0'$, so

$$c^2 = (-\rho_0') \cdot \frac{H}{j^2\pi^2} \cdot gH.$$

If we write $\rho_0' \approx (\rho_{top} - \rho_{bot})/(H\bar{\rho})$ we get

$$c^2 = \frac{1}{j^2\pi^2} \cdot g'H, \quad g' = g\frac{\rho_{\text{bot}} - \rho_{\text{top}}}{\bar{\rho}}$$

$$\approx \text{ reduced gravity.}$$

In this case, we get a whole family of waves, each member of the family having different vertical structure. As the vertical structure becomes more complex, i.e., as j increases, the waves go slower.

In the flat-bottom case, the modes are independent, and each equilibrates at its own speed.

The free-surface case is a bit more complicated. If the position of the free surface is given by $z = \eta(x, y, t)$, the free-surface boundary condition is $\eta_t = w(x, y, 0, t) = W(x, y, t)F(0)$. The evolution equation for η is derived by integrating the continuity equation to form

$$\eta_t + \int_{-H}^{0} u_x + v_y \, \mathrm{d}z = 0. \tag{4.71}$$

The pressure in (4.53)–(4.54) is given by

$$p = g\eta + \int_{z}^{0} g\rho \, \mathrm{d}z. \tag{4.72}$$

Substituting (4.61) into the above integral leads to

$$\eta + RF'(0) = 0. \tag{4.73}$$

Differentiating (4.73) with respect to time and substituting the free-surface boundary condition for η_t and then applying (4.61) we find

$$W(0)\left(F(0) - \rho_0'(0)\frac{F'(0)}{F''(0)}F'(0)\right) = 0. \tag{4.74}$$

Equation (4.74) then becomes

$$F(0) - (c^2/g)F'(0) = 0, \tag{4.75}$$

by virtue of (4.67). This is the upper boundary condition for the vertical modes in the case of free-surface boundary conditions. As in the rigid lid case, this determines the speed of the waves, but the speed is now the solution to a transcendental equation that cannot be written down directly. The surface pressure can be identified directly with (4.63).

In the case of constant N, $F(z) = \sin(N(z+H)/c)$, so (4.75) becomes

$$\tan(NH/c) = Nc/g. \tag{4.76}$$

What do we expect the solutions of (4.76) to look like? If $NH/(gH)^{1/2} << 1$, we find $c \approx \pm(gH)^{1/2}$, so this system admits the surface gravity wave that was suppressed in the rigid lid case. For c so small that we can consider $NHc/g \approx 0$, we recover the rigid lid case $NH/c = j\pi$ for nonzero integer j. The surface pressure can be identified directly with (4.72), which can, in turn, be identified with the formally identical expression for surface pressure in the rigid lid.

We have seen that higher mode waves, i.e., those with more complex vertical structure, travel more slowly than their lower mode counterparts. Let us take a moment to examine the implications of this fact for spinup of numerical models. Adjustment of a model ocean to changes in the forcing is mediated in large part by waves of the sort described above, and one rough estimate of the adjustment time is the scale of the model domain divided by the speed of the slowest wave that participates in the adjustment. Higher mode waves can be excited by interaction of lower mode motions with topography. We therefore expect models with realistic topography to adjust to changes in forcing more slowly than flat-bottom models. Since it is impractical to calculate the fully adjusted state of a general circulation model (GCM) to given forcing fields – and it is quite possible that, for some models, no fully adjusted stable steady state exists – we expect any given model to undergo a significant adjustment process, beginning at the initial time. In many cases the model state during this adjustment period may not correspond to any state likely to be observed in the real ocean. By the foregoing, in a reasonably detailed model with detailed ocean bottom topography and fine vertical resolution the adjustment process could take quite a long time, in general much longer than the adjustment time required by a flat-bottom model, or one with coarse vertical resolution. As the vertical resolution increases, the model will admit more and more vertical modes with slower and slower timescales, and thus one expects such models to take longer and longer to settle down.

One illustration of this was presented by Anderson *et al.* (1979). They ran a linearized version of the Bryan–Cox model (Bryan (1969)), driven by idealized winds. Their implementation had 1° resolution in the horizontal and 12 vertical levels. Figure 4.2, taken from Anderson *et al.* (1979) shows the bathymetry in meters they used in their simulations.

Figure 4.3 shows the evolution of streamfunction sections at 30° N in the flat- and rough-bottom cases. By day 500, the flat-bottom case has reached steady state, while the rough-bottom case has not reached steady state in 1500 days. Even in this relatively simple linear case, adjustment of the total transport to imposed wind forcing in a model with significant bottom topography can be very much slower than it would be in the flat-bottom case.

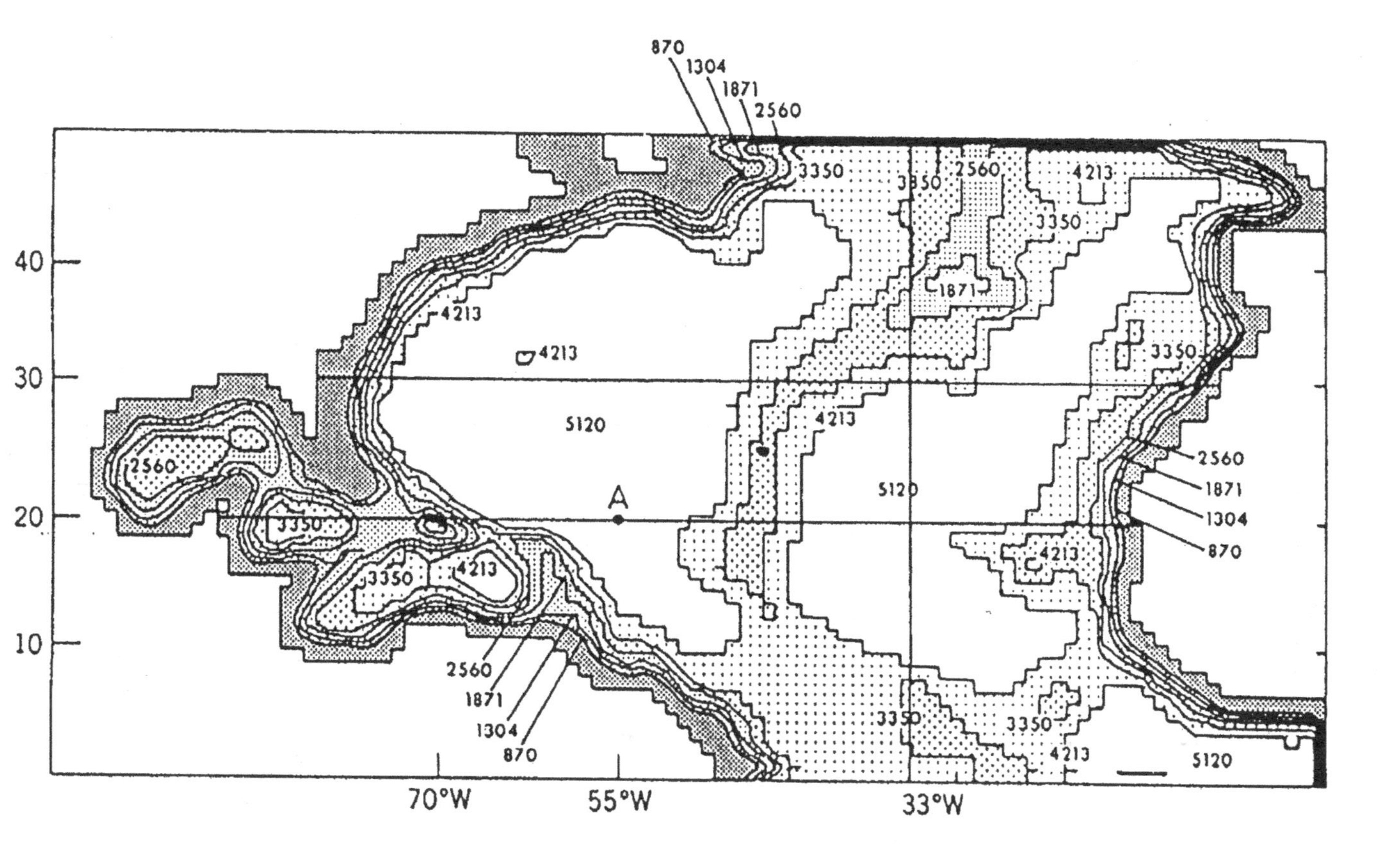

Fig. 4.2: Coarsely resolved topography of the North Atlantic used in the primitive equation spinup experiments of Anderson *et al.* (1979). Reproduced from Figure 4 of Anderson *et al.* (1979), with permission of the American Geophysical Union.

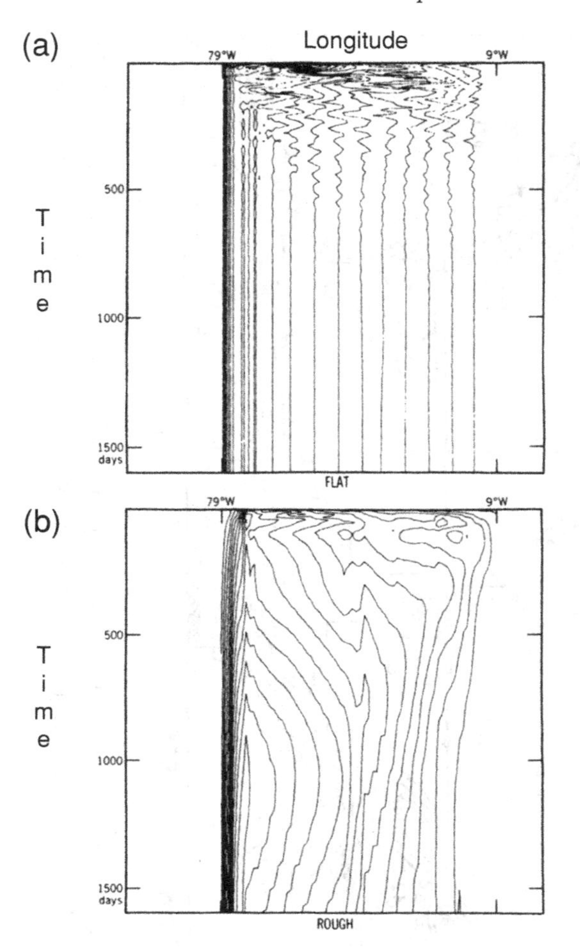

Fig. 4.3 Contour plots of transport streamfunction in the longitude–time plane of the transport streamfunction along 30° N from the primitive equation spinup experiments of Anderson *et al.* (1979). (a) Results from a spinup experiment with a flat bottom. (b) Results of a similar spinup experiment with bottom topography as shown in Figure 4.2. Reproduced from Figure 14 of Anderson *et al.* (1979), with permission of the American Geophysical Union.

4.5 Consequences of discretization

The lateral and vertical grids used in the Bryan–Cox model are shown in Figures 4.4 and 4.5, taken from Cox (1984). The horizontal grid is the Arakawa B-grid. This was the original choice for the model. Given

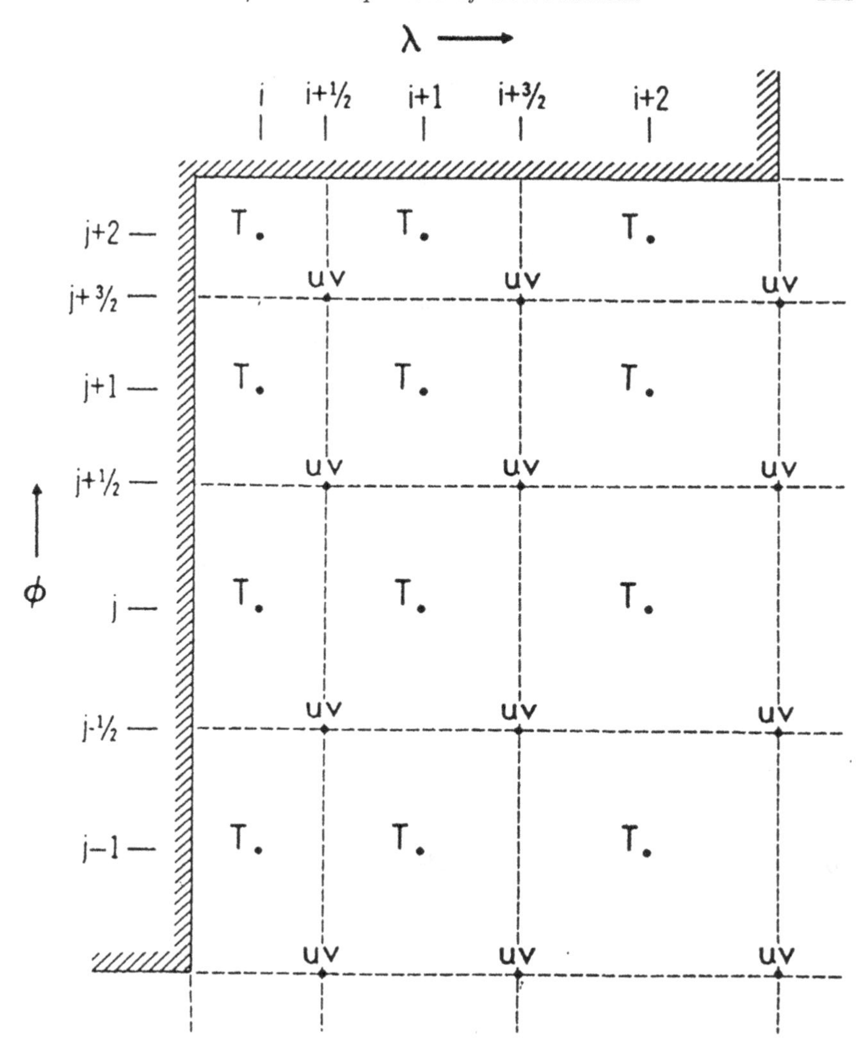

Fig. 4.4 Horizontal grid layout for the Bryan–Cox model. Reproduced from page C4 of Cox (1984), with permission of Geophysical Fluid Dynamics Laboratory, NOAA.

the original intent to study basin-scale models with coarse resolution, this appears to have been a good choice; we have seen that the B-grid is superior to the C-grid when the waves are not well resolved.

Vertical discretization in terms of the height z, however, does not result in efficient representation of the vertical mixing process. This is

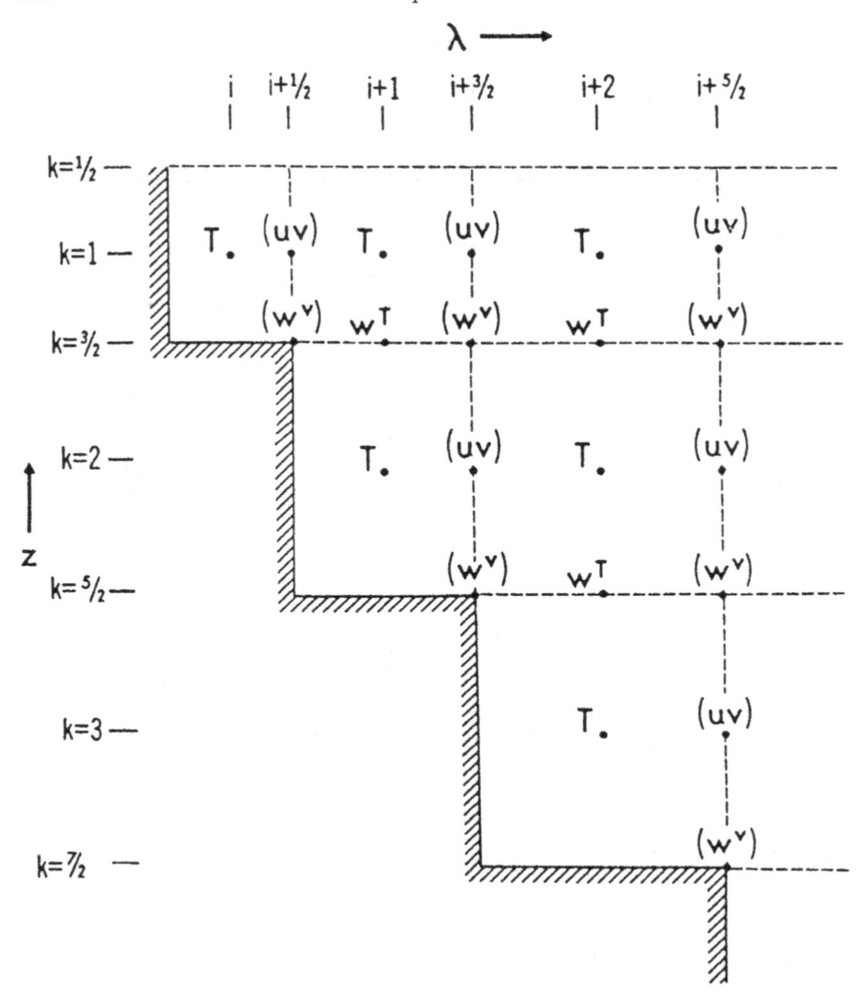

Fig. 4.5 Vertical grid layout for the Bryan–Cox model. Reproduced from page C4 of Cox (1984), with permission of Geophysical Fluid Dynamics Laboratory, NOAA.

because diffusion along isopycnal surfaces is much stronger than diffusion across them, and if the isopycnal surfaces have significant slope, the vertical mixing process will be poorly represented. Redi (1982) noted that, in studies by Sarmiento (1982), the Bryan–Cox model tended to underpredict tritium concentrations below 450 m in the North Atlantic. Typically, the vertical diffusion coefficients are much smaller than the horizontal ones. In this case, diffusion along sloping density surfaces could be significantly understated; this is consistent with Sarmiento's results. Redi (1982) suggested replacing the diffusion coefficients A_{H} and A_{V} in the tracer equations with a tensor $\mathbf{K}$, so the equations for tracers would become

$$\theta_t + u\theta_x + v\theta_y + w\theta_z = \nabla \cdot \mathbf{K}\nabla\theta,$$
$$s_t + us_x + vs_y + ws_z = \nabla \cdot \mathbf{K}\nabla s.$$

The simple z-coordinate model can be written in this way, with

$$\mathbf{K} = \begin{pmatrix} 1 & 0 & 0 \\ 0 & 1 & 0 \\ 0 & 0 & \epsilon \end{pmatrix}, \tag{4.77}$$

where the factor ϵ in the $(3,3)$ element of $\mathbf{K}$ is an attempt to represent the fact that diffusion is generally stronger in horizontal than in vertical directions. In the cited tracer studies (e.g., Sarmiento and Bryan, 1982), the ratio ϵ of the vertical to the horizontal diffusivities is 10^{-7}. Redi (1982) pointed out that this form of the diffusion tensor is, in fact, more appropriate for the system in which the third direction is normal to the local isopycnal surface. She therefore suggested that the local isopycnal slope be determined and a suitable rotation matrix be calculated to determine the form of $\mathbf{K}$ in z-coordinates when $\mathbf{K}$ takes the form (4.77) in isopycnal coordinates. If the isopycnals have significant slope, diffusive fluxes along isopycnal surfaces can result in tracers spreading vertically. As a concrete example, suppose the density is given locally by $\rho = \rho_0(z - \alpha x)$. If the diffusion tensor takes the form (4.77), with the first two directions tangential to the local isopycnal and the third direction normal to it, then the result of transforming the diffusion tensor back to the x–y–z coordinate system is

$$\mathbf{K} = \begin{pmatrix} 1 + \epsilon\alpha^2 & 0 & (1-\epsilon)\alpha \\ 0 & 1 + \alpha^2 & 0 \\ (1-\epsilon)\alpha & 0 & \epsilon + \alpha^2 \end{pmatrix}, \tag{4.78}$$

Geometrically, α is the sine of the angle between the normal to the local isopycnal surface and the vertical. From the form of (4.78), it is clear that even if there is no cross isopycnal diffusion at all, diffusion along sloping isopycnals will give rise to vertical diffusive fluxes. Redi (1982) gives detailed if indirect evidence to support the hypothesis that the underprediction of tritium concentrations in the North Atlantic is due to this problem. In much of the North Atlantic, at selected depths between 25 m and 713 m, $\alpha^2 > 5 \times 10^{-7}$, and in those regions most of the vertical diffusive fluxes are tangent to sloping isopycnals.

Cox (1987) presented several simple examples of the process, along with a computational experiment in which results of a simulation in which adjustment of the diffusion tensor to be diagonal with respect to an isopycnal coordinate system were compared with an earlier simulation in which the simple vertical mixing scheme was used. He found that sharper frontal structures and more accurate deep and bottom water formation is produced, as expected. Current versions of the model contain a variety of options for anisotropic diffusion. There has been extensive development of methods along this line since the work of Redi and Cox; see, e.g., Wajsowicz (1993), Griffies (1998) and Griffies *et al.* (1998).

4.6 The importance of vertical resolution

Weaver and Sarachik (1990) used the Bryan–Cox model to produce a striking illustration of the importance of vertical resolution in models of this type. In that study they examined a meridional overturning cell that had been observed to occur near the equator in a number of modeling studies. An example can be found in Bryan (1987). Bryan found a deep overturning cell near the equator, whose mass transport decreased with increasing vertical diffusivity. The behavior of this overturning cell is illustrated in Bryan's Figure 7, shown here as Figure 4.6.

The quantity plotted in Figure 4.6 is the meridional overturning streamfunction. This quantity is commonly used to represent the thermohaline circulation. It is derived by integrating the continuity equation with respect to longitude. Write the continuity equation in spherical coordinates:

$$u_\lambda + (v \cos \phi)_\phi + a \cos \phi w_z = 0, \tag{4.79}$$

where λ is the longitude, ϕ is the latitude and a is the radius of the Earth. Integrating from the western boundary λ_w to the eastern boundary λ_e leads to

$$\int_{\lambda_w}^{\lambda_e} (v \cos \phi)_\phi \, \mathrm{d}\lambda + a \cos \phi \int_{\lambda_w}^{\lambda_e} w_z \, \mathrm{d}\lambda = 0, \tag{4.80}$$

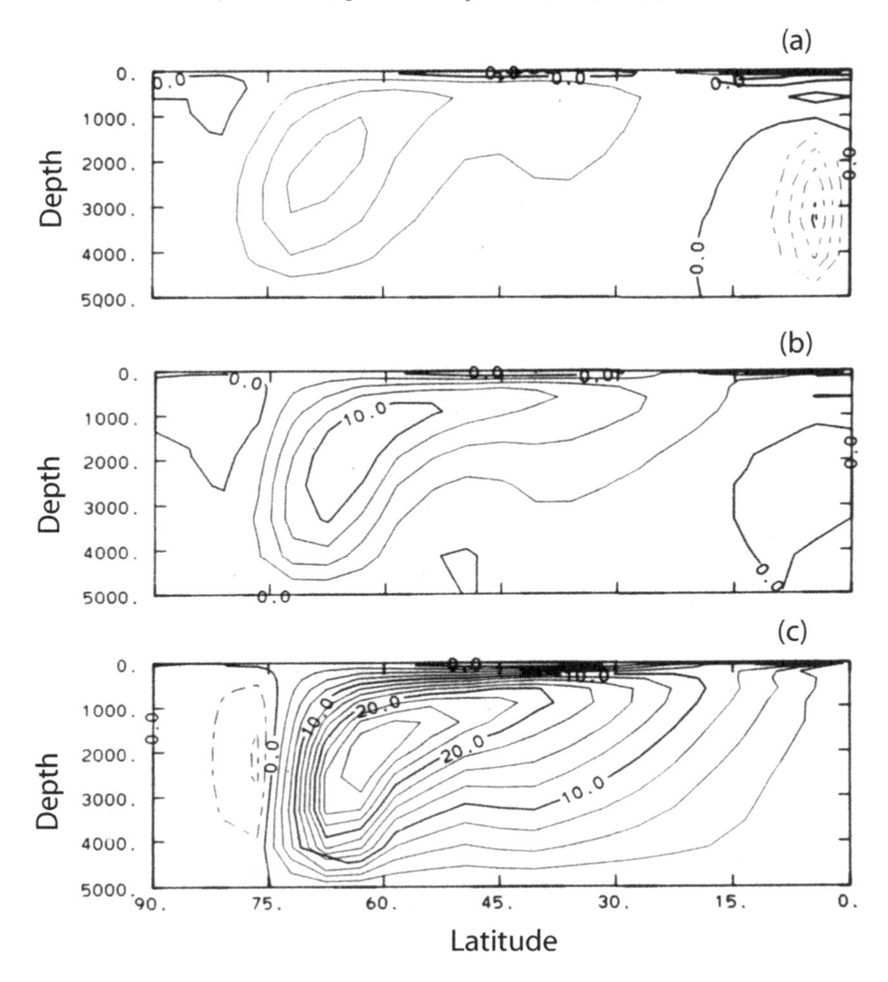

Fig. 4.6 Meridional overturning streamfunction in three different simulations. (a) $A_{HV} = 0.1\,\mathrm{cm^2\,s^{-1}}$; (b) $A_{HV} = 0.5$; (c) $A_{HV} = 2.5$. Contour interval $= 2.5 \times 10^6\,\mathrm{m^3\,s^{-1}}$, solid contours indicate clockwise circulation. Reproduced from Figure 7 of Bryan (1987), with permission of the American Meteorological Society.

since the first term vanished because of the boundary condition. From the integrals in (4.80) we may define a streamfunction Φ by

$$-\Phi_z = \int_{\lambda_w}^{\lambda_e} (v \cos\phi)\, \mathrm{d}\lambda, \tag{4.81}$$

$$\frac{1}{a}\Phi_\phi = \int_{\lambda_w}^{\lambda_e} wa \cos\phi\, \mathrm{d}\lambda. \tag{4.82}$$

Bryan (1987) notes that the deep overturning cell is "primarily due to downwelling on the eastern boundary at the equator. The physical processes driving this circulation are difficult to determine ..." and cites earlier studies in which similar features were observed.

In Weaver and Sarachik (1990), the vertical resolution in a model grid with horizontal grid spacing of 2°, covering a 60°-wide southern hemisphere basin extending from 70° S to the equator was varied. Their model ocean had uniform depth of 5000 m. The vertical eddy viscosity was set to $1.0 \times 10^{-4}\,\mathrm{cm^2\,s^{-1}}$ and the vertical heat diffusivity varied from $0.3 \times 10^{-4}\,\mathrm{cm^2\,s^{-1}}$ at the surface to $1.3 \times 10^{-4}\,\mathrm{cm^2\,s^{-1}}$ at the bottom. Horizontal eddy viscosity and diffusivity were set to $5 \times 10^4\,\mathrm{m^2\,s^{-1}}$ and $1 \times 10^3\,\mathrm{m^2\,s^{-1}}$ respectively. A symmetric boundary condition was applied at the equator. Three experiments were performed, one with 12 levels in the vertical, chosen to be identical to the vertical discretization in a number of earlier studies (Bryan, 1987; Toggweiler *et al.*, 1989, and references therein), one with 19 levels and one with 33 levels. In the experiment with 12 levels, the overturning cell evident in Figure 4.6 is evident; compare that figure to Figure 4.7. In order to eliminate the possibility that the major features of the meridional circulation were sensitive to the symmetric boundary condition at the equator, an experiment was performed in which a basin extending from 70° S to 70° N was initialized with symmetric conditions and integrated for a further 20 000 time steps. The deep overturning cell near the equator persisted.

In the experiment with 19 vertical levels, the model was initialized by interpolation from the end result of the experiment with 12 vertical levels and continued until a steady state was reached. Figure 4.8 shows that the spurious downwelling and resulting counter-rotating deep cell near the equator is replaced by a weaker deep cell of opposite sense, but the origin of this deep cell is not immediately apparent. If the resolution is increased still further, the cell disappears entirely, as shown in Figure 4.9. Weaver and Sarachik argue that the artificial cells arise as a consequence of the oscillatory behavior associated with excessive

Table 4.1 Summary of Reynolds and Peclet numbers in experiments described by Weaver and Sarachik. Reynolds and Peclet numbers are calculated relative to vertical velocities and grid spacings

Experiment number	Levels	Figure reference	Cell Reynolds number	Cell Peclet number
1	12	4.7	3.2	4.0
2	19	4.8	1.6	2.0
3	33	4.9	1.0	1.3

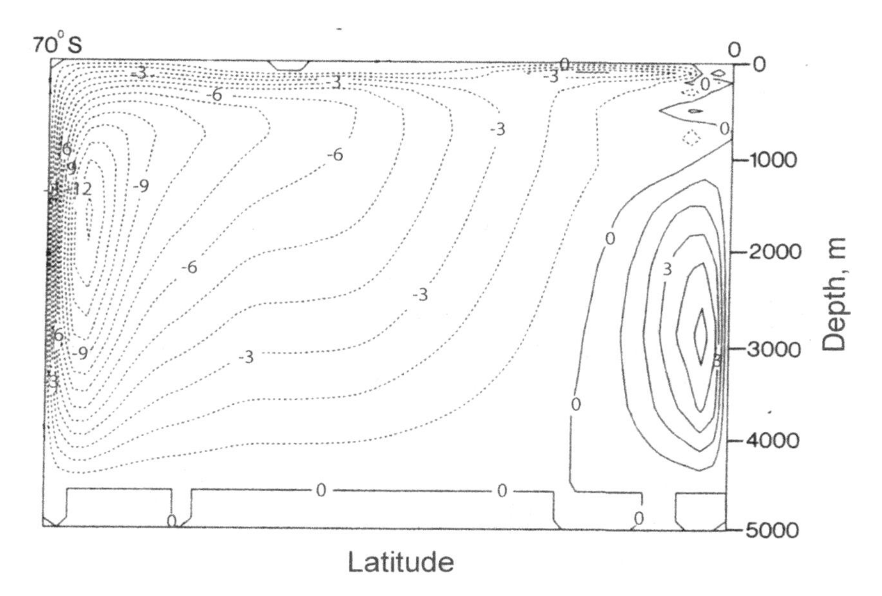

Fig. 4.7 Meridional overturning streamfunction for the southern ocean basin experiment with 12 vertical levels, after integration for 280 000 time steps, corresponding to 2628 years at the uppermost level and 7885 years at the lowest level. Reproduced from Figure 2 of Weaver and Sarachik (1990), with permission of the American Meteorological Society. Note the overturning cell near the equator.

Peclet and Reynolds numbers as described in Section 3.3.1, and they present a calculation similar to the one shown in that section. Calculated Reynolds and Peclet numbers from Weaver and Sarachik's experiments are shown in Table 4.1. The results displayed in Weaver and Sarachik (1990) provide a clear illustration of the importance of vertical resolution in ocean models in which centered schemes are used in the vertical.

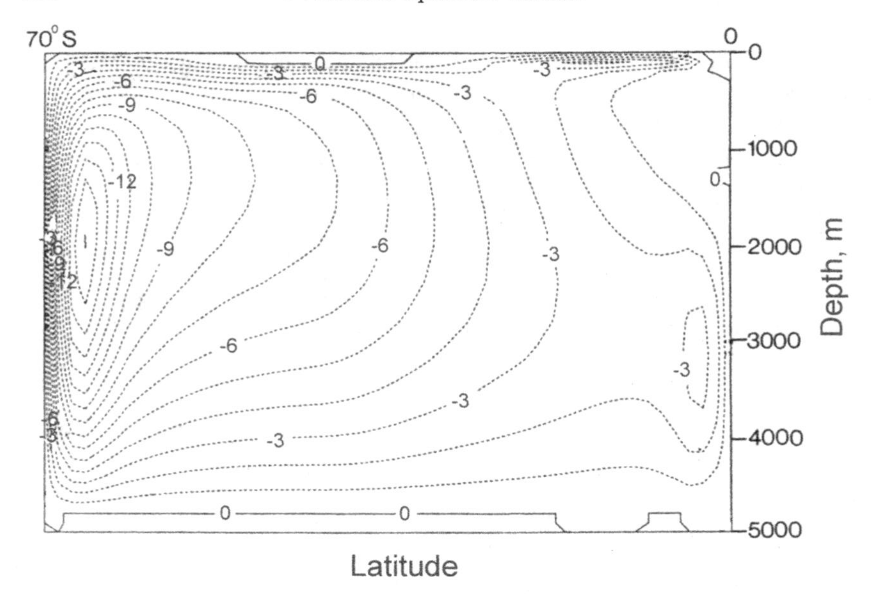

Fig. 4.8 Meridional overturning streamfunction for the southern ocean basin experiment with 19 vertical levels, after integration for 440 000 time steps, corresponding to 4381 years at the uppermost level and 13 142 years at the lowest level. Reproduced from Figure 3 of Weaver and Sarachik (1990), with permission of the American Meteorological Society.

4.7 Example: Transport in the Drake Passage and the large-scale circulation

In this section we examine the use of the Bryan–Cox model to investigate the influences on the formation of water-masses. In the study of water-mass formation, a model can be particularly valuable. Subsurface data are sparse and their distribution is far from uniform, and some quantities of interest such as subduction rates are extremely difficult to measure directly. Some of the processes of interest are also very slow; changes in large-scale ocean circulation that accompany climate change may take place on timescales of centuries. Analytical studies are valuable, but necessarily highly idealized.

Models allow us to perform conceptual experiments; we may, for example, gain insight into the role of the Antarctic Circumpolar Current (ACC) in the large-scale circulation by performing model simulations in which the Drake Passage is closed off and the circumpolar flow is blocked.

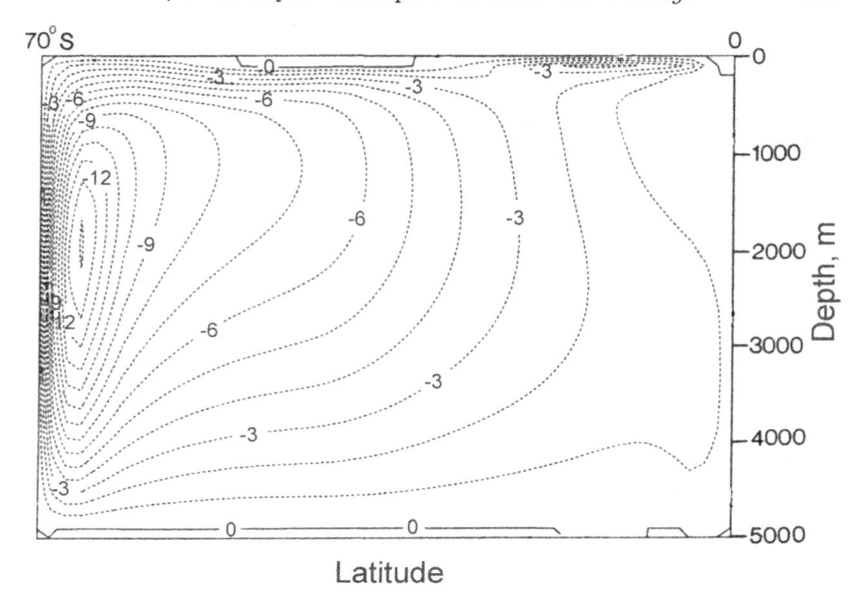

Fig. 4.9 Meridional overturning streamfunction for the southern ocean basin experiment with 33 vertical levels, after integration for 600 000 time steps, corresponding to 6133 years at the uppermost level and 18 398 years at the lowest level. Reproduced from Figure 4 of Weaver and Sarachik (1990), with permission of the American Meteorological Society.

Model studies with domains of global extent and with grid resolutions sufficiently fine to resolve the mesoscale eddies are highly resource intensive, and are, in general, too expensive to use for model simulations longer than a few decades. Such model runs are useful for study of the wind-driven circulation above the thermocline, but after a few decades of integration, the convergence of the deep flow to statistical equilibrium is not complete. Cox (1989) notes some studies in which modelers tried to surmount this problem by the "robust diagnostic technique," in which the model is forced to relax toward observed data in the deep ocean. Cox notes that the results of these experiments have been equivocal.

Problems with the robust diagnostics method are related to the fact that the steady state of the model may differ significantly from observed data. It has long been known to the numerical weather prediction community that the long-term average of the output of weather prediction models differs significantly from natural climate, and the tendency of

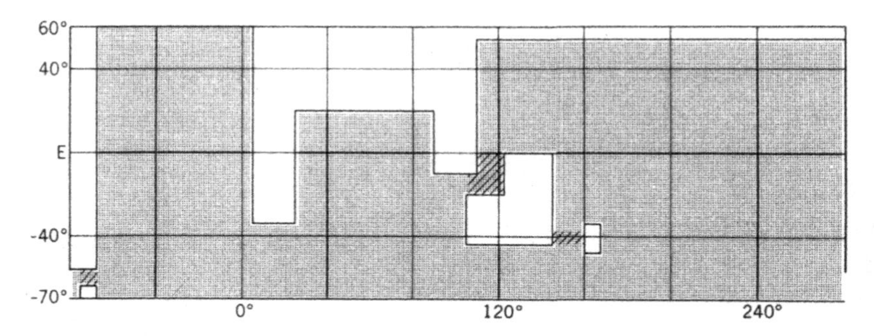

Fig. 4.10 The region covered by the model. The hatched areas designate sills, as noted in the text. The domain is periodic in the zonal direction. Reproduced from Figure 1 of Cox (1989), with permission of the American Meteorological Society.

atmospheric models to approach an unrealistic mean state is known as "climate drift." Use of the robust diagnostics method compounds the problem of climate drift in ocean models. The data inserted in the process of implementing the robust diagnostics method are usually not the direct result of observations, but rather the result of producing maps of ocean variables on a grid by interpolation of archived observations. These gridded data sets are themselves subject to appreciable errors in data-sparse regions.

Cox (1989) avoided the need to apply the robust diagnostics method by setting up the model with low resolution, so the model could be run for a very long time. He contended that this approach does less violence to the integrity of solutions obtained for the sub-thermocline ocean, and predicted that long runs with eddy-resolving models were a decade in the future. As this text is being written, Cox's decade is up, and extensive long-term studies with eddy-resolving global circulation models have begun to appear; see Section 4.12 for a few examples.

In this study, the Bryan–Cox model was set up with zonal resolution of 2.5°, meridional resolution of 2° and 12 vertical levels over most of the ocean. Midocean ridges are neglected and shallow sills are placed between Australia and New Zealand, and between Australia and Asia. There is also a shallow sill at the Drake Passage in those experiments in which the passage is left open. The model geometry is depicted in Figure 4.10.

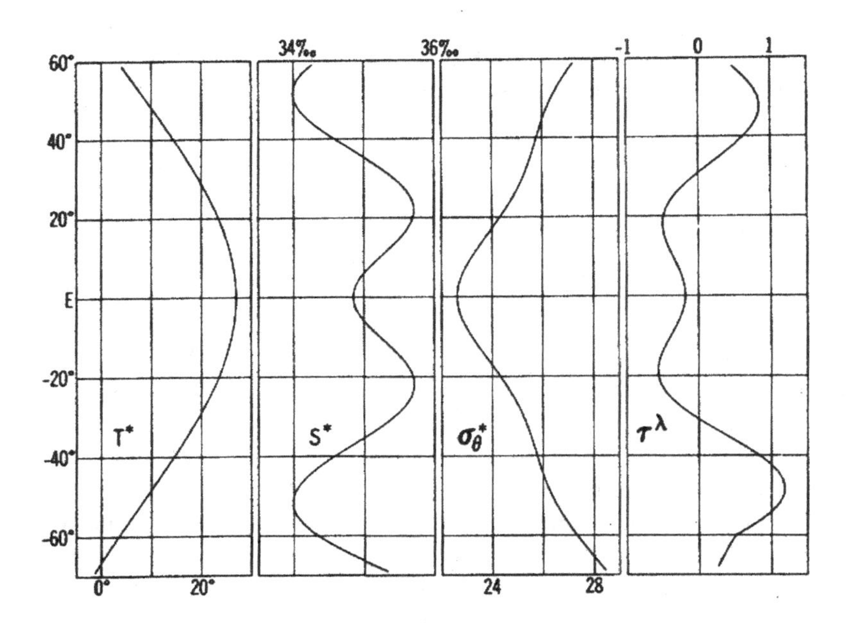

Fig. 4.11 Surface boundary conditions of the model. σ_θ is calculated directly from T^* and S^*. The wind stress is given in units of dyn $\cdot$ cm^{-2}. Reproduced from Figure 2 of Cox (1989), with permission of the American Meteorological Society.

The horizontal eddy viscosity is taken to be $4 \times 10^4\,\mathrm{m^2\,s^{-1}}$ and the vertical eddy viscosity is $5\times10^{-3}\,\mathrm{m^2\,s^{-1}}$. Cox notes that this latter value is too high, but necessary to suppress noise in equatorial regions. The ratio of these values of the horizontal and vertical viscosity is in line with that noted by Redi (1982). The horizontal diffusivity of heat and tracers is $1 \times 10^3\,\mathrm{m^2\,s^{-1}}$ and the vertical diffusivity is $5 \times 10^{-5}\,\mathrm{m^2\,s^{-1}}$. With these parameter choices, we expect these simulations to be susceptible to the problems described by Redi (1982) and by Weaver and Sarachik (1990).

Surface boundary conditions are shown in Figure 4.11. These correspond roughly to long-term mean observational data for temperature, salinity and wind stress at the surface of the real ocean. The meridional component of wind stress is neglected. Note that at the extreme south the surface water is colder and saltier than it is in the North.

The presence of the ACC plays a fundamental role in controlling the relative rate of deep and bottom water formation between the northern and southern hemispheres. The density of the southern boundary of the ACC is determined by surface forcing, and we expect the eastward-flowing ACC to be balanced geostrophically, at least in part. This corresponds to a barrier of relatively low density water across the ACC, which should tend to isolate the dense water of the extreme southern ocean to the water north of the ACC.

In an effort to isolate the effect of the ACC, two runs were performed on a domain with a single basin. The domain for experiments with a closed basin was constructed by blocking the Drake Passage and extending the schematic boundary of west Africa southward to the boundary. In the other experiment, a gap was allowed at the Drake Passage and another gap at the eastern boundary. Periodic boundary conditions were imposed at the gaps.

Zonally averaged temperature and salinity profiles for model runs with and without a gap are shown in Figures 4.12 and 4.13. In the run with the gap, the density gradients near 60° S are greater, corresponding to the geostrophically balanced component of the ACC. Figure 4.12 shows that in the case with the gap, the subthermocline water is significantly warmer. The salinity map shown in Figure 4.13 indicates that the influence of high salinity surface waters prescribed at the southern boundary decreases in the interior due to the presence of the throughflow. This is accompanied by a tongue of low salinity water originating from the surface salinity boundary condition at the northern wall and penetrating southward at intermediate depth. This might be considered to be analogous to North Atlantic Deep Water in the real ocean, but the salinity signature is of opposite sign.

The meridional overturning streamfunction for these experiments is shown in Figure 4.14. In Figure 4.14(a), representing the non-gap case, the circulation is dominated by a single overturning cell. Water sinks in a narrow region adjacent to the southern boundary and upwells over most of the domain, with some of this water appearing north of 50° N at a depth of about 1500 m. Subsurface water formation near the northern boundary is confined to a shallow cell.

When the gap is present, there are three overturning cells (Figure 4.14(b)). An overturning cell appears between the southern boundary and the gap. This cell rotates in the opposite sense from the dominant cell in the experiment with no gap. Most of the domain is occupied by a weaker overturning cell, in which subsurface water is produced just *north* of the gap. The cell representing the production of subsurface water in the north is stronger and deeper than the

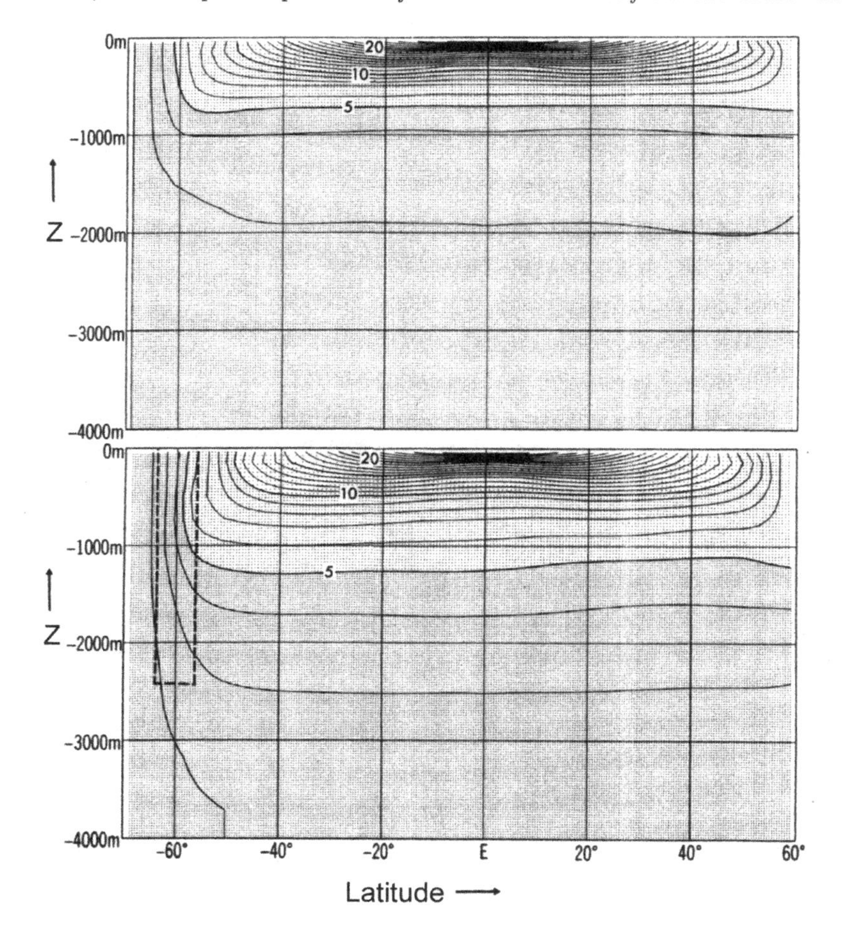

Fig. 4.12 Zonally averaged temperature in the single-basin experiments with no gap (top) and with a gap (bottom) as designated by the dashed line. Reproduced from Figure 3 of Cox (1989), with permission of the American Meteorological Society.

corresponding cell in the experiment with no gap, but does not extend much further south.

4.8 Example: Separation of the Brazil current from the coast

In this section we present an example of a study in which the Bryan–Cox model was used to investigate the question of why the Brazil current separates from the coast where it does. This example follows the work

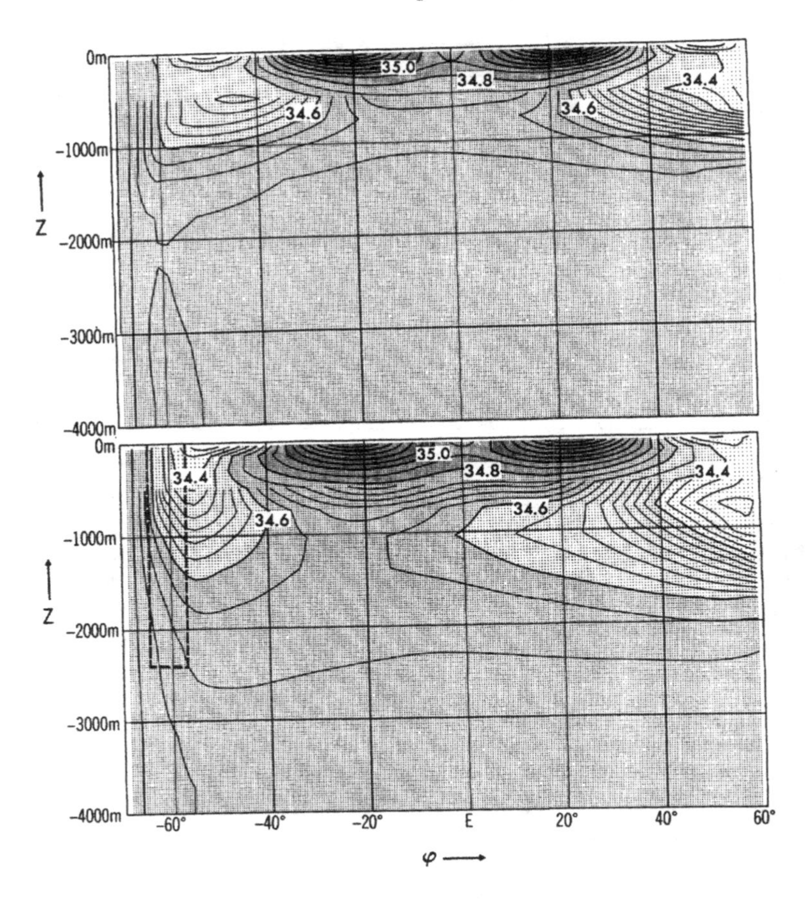

Fig. 4.13 As in Figure 4.12, but for zonally averaged salinity. Reproduced from Figure 4 of Cox (1989), with permission of the American Meteorological Society.

of Matano (1993). The Brazil current flows southward along the continental slope of South America to its separation point near 38° S. It is a fairly weak current as western boundary currents go, with a total transport that may not exceed 20 Sv (Gordon and Greengrove, 1986). The simple viscous theory (Stommel, 1966) predicts separation at the latitude at which the wind-stress curl vanishes. At issue here is the fact that the Brazil current separates *north* of the latitude of the zero of the wind-stress curl. Modifications to the theory that take inertial effects into account (Stommel, 1966, Chapter 8) would be expected to predict the Brazil current to overshoot the zero of the wind-stress curl,

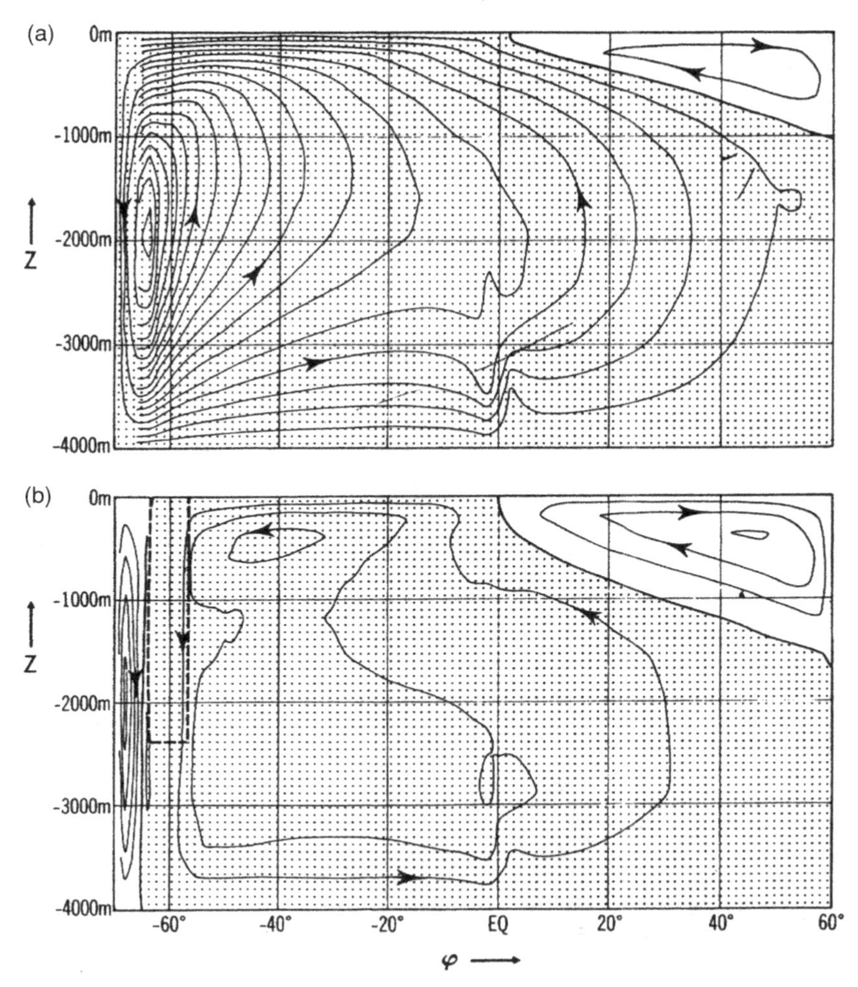

Fig. 4.14 Meridional mass transport streamfunction in the single-basin experiments with no gap (top) and with a gap (bottom) as designated by the dashed line. Each contour represents 2×10^6 m^3 s^{-1} of flow. Reproduced from Figure 5 of Cox (1989), with permission of the American Meteorological Society.

i.e., one would expect an inertial theory to predict the separation point *south* of the zero of the wind-stress curl. Topographic steering is another possibility, but Matano points out that the continental slope of South America is relatively smooth compared to other western boundaries. In this work, Matano set out to use the model to decide among different possible explanations of the separation latitude.

He began with a simple analytical theory in which input of vorticity by the wind-stress curl is balanced by Rayleigh friction and the β effect. The resulting equation for the transport streamfunction ψ is then

$$\epsilon \nabla^2 \psi + \frac{\partial \psi}{\partial x} = \mathbf{k} \cdot \nabla \times \tau. \tag{4.83}$$

The boundary condition for ψ was chosen to reflect the inflow of the Antarctic Circumpolar Current (ACC) through the Drake Passage at the western boundary, balanced by a combination of inflow and outflow at the eastern boundary representing the ACC and the Agulhas current. The first task is to investigate the relative importance of boundary forcing and wind forcing in this simple model. First, τ is set to zero in (4.83). The result is shown in Figure 4.15.

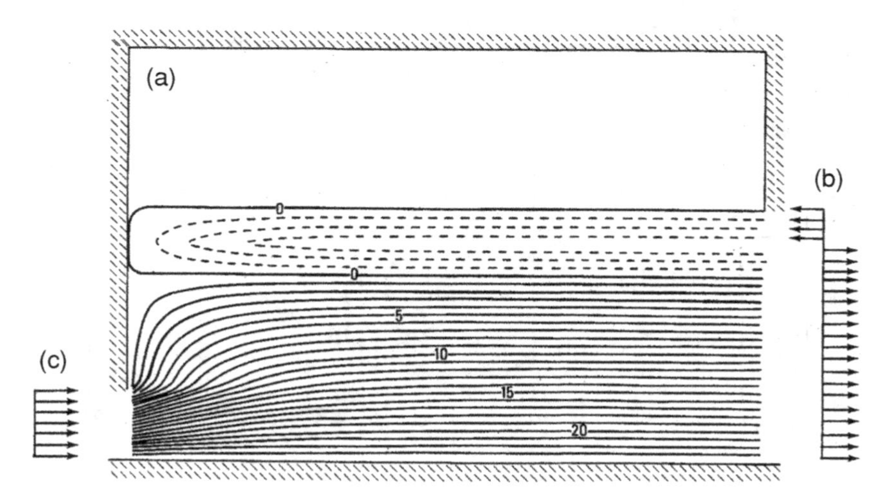

Fig. 4.15 (a) The solution to (4.83) in an idealized South Atlantic domain with no wind forcing. (b) The boundary condition imposed at the eastern boundary. (c) The open-boundary conditions imposed at the western boundary. Reproduced from Figure 1 of Matano (1993), with permission of the American Meteorological Society.

Since there are no vorticity sources in the solution without wind stress, the inflow at the east forms a current that flows due west until it meets the western boundary where it forms the analog of the Brazil current in this model. It follows the coastline southward to its confluence with the northward-flowing analog of the Malvinas current, where it separates from the coast and joins the broad eastward flow of the ACC. The separation point is entirely a function of the boundary conditions, which represent parameterizations of the effect of wind stress in the Pacific and Indian Oceans.

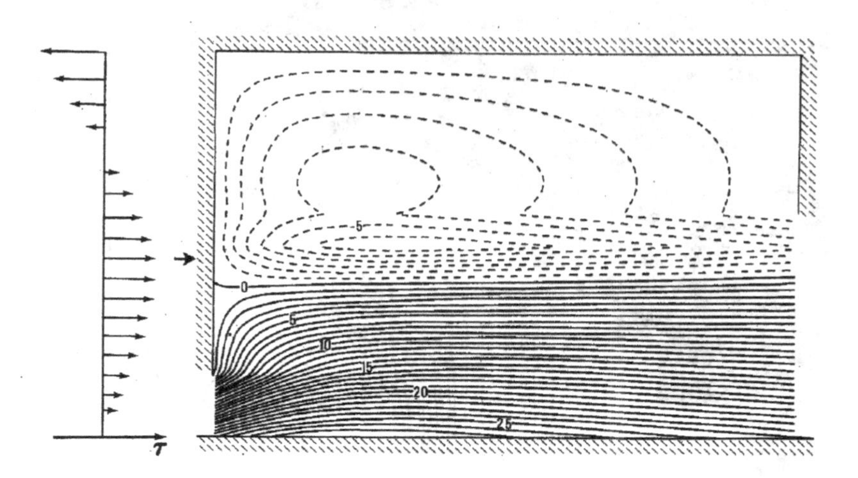

Fig. 4.16 The solution to (4.83), similar to Figure 4.15, but with the addition of wind forcing, as shown at the left. The arrow shows the latitude at which the western boundary current separates in Figure 4.15. The east–west tilt of the zero streamline is the result of the competing effects of the wind-stress forcing and the open-boundary condition. The arrow shows the latitude of separation in the case with no wind-stress curl shown in Figure 4.15. Reproduced from Figure 2 of Matano (1993), with permission of the American Meteorological Society.

The result of adding a simple wind pattern is shown in Figure 4.16. A broad subtropical gyre has appeared as a result of the wind-stress curl, and separation occurs at the zero of the wind-stress curl, south of the point at which it occurs in the previous case.

This simple model makes plausible the possibility that the northward flowing Malvinas current can result from wind-stress curl, either directly acting on the South Atlantic, or on the Pacific or Indian Oceans through the ACC and the Agulhas current. It could also result from a portion of the ACC following f/H contours from the Drake Passage northward along the Argentinian coast. A map of these f/H contours is shown in Figure 4.17.

The domain and grid used by Matano is shown in Figure 4.18. Artificial solid boundaries are imposed at the northern and southern

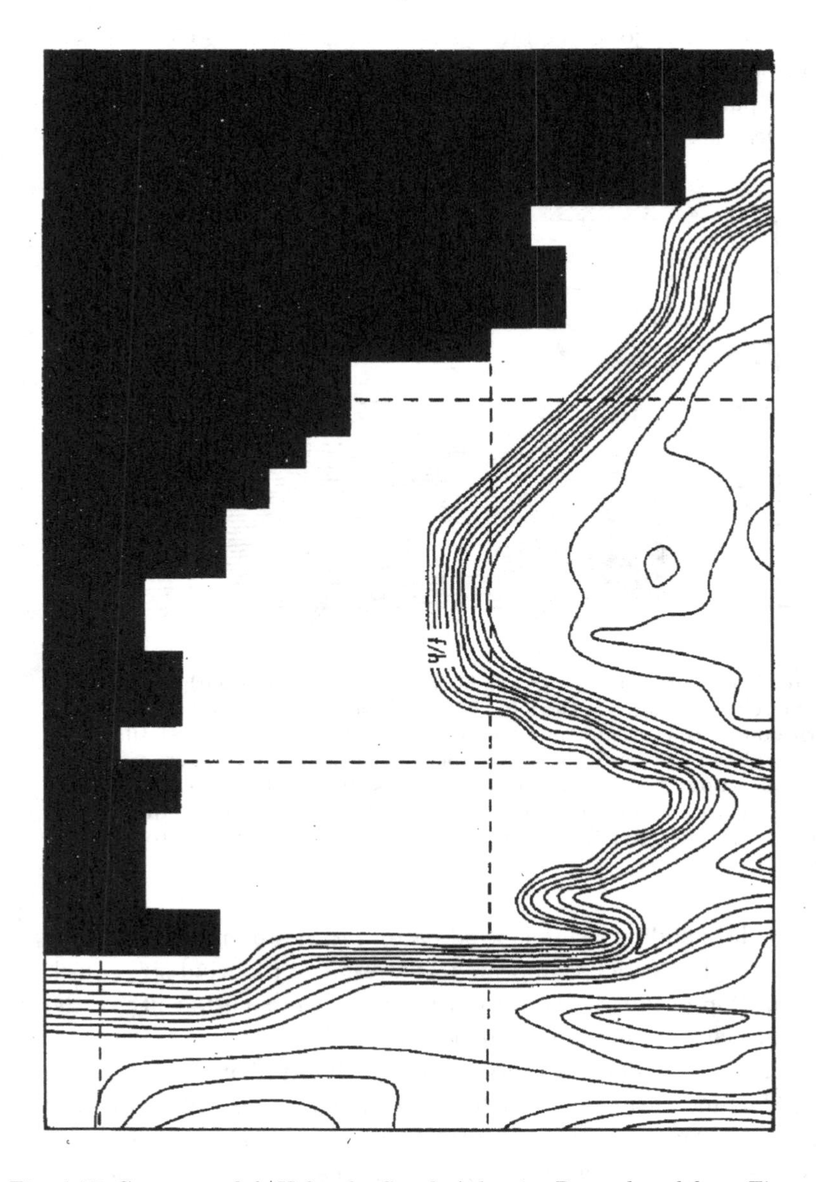

Fig. 4.17 Contours of f/H for the South Atlantic. Reproduced from Figure 4 of Matano (1993), with permission of the American Meteorological Society.

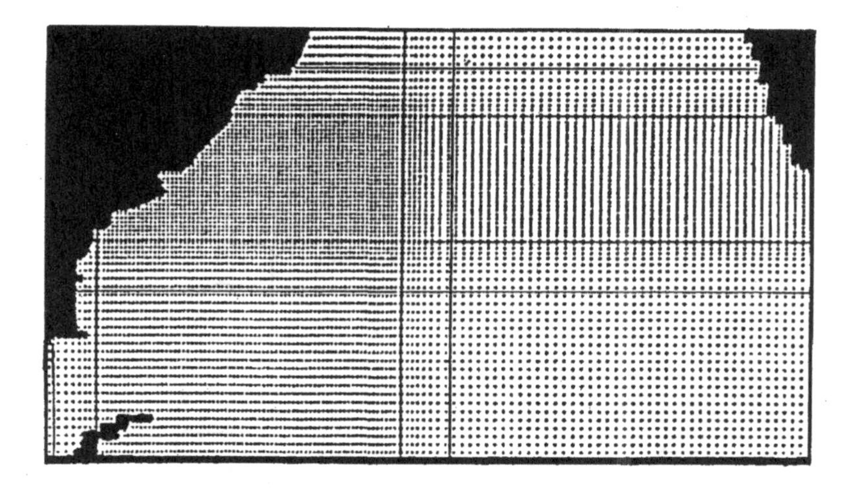

Fig. 4.18 Domain and grid of the model used by Matano to investigate the separation of the Brazil current. Reproduced from Figure 5 of Matano (1993), with permission of the American Meteorological Society.

edges of the domain. The transport streamfunction is prescribed at the boundaries. The transport profiles imposed across the open boundaries in the basic experiment are shown in Figure 4.19. These conditions were varied, as described here, to investigate the sensitivity to the boundary conditions.

At the inflow points, temperature and salinity are prescribed, and the baroclinic velocities are geostrophically adjusted. A radiation condition is imposed at outflow points.

The wind stress applied was an analytical approximation to the zonally-averaged mean wind stress of Hellerman and Rosenstein (1983). The basic analytic form used with this model was

$$\tau^x(y) = 0.2 + \tanh\left[\frac{(y+90)}{20}\right], \quad y > 45^\circ \text{ S}, \tag{4.84}$$

$$\tau^x(y) = -0.2 + \tanh\left[\frac{(y+30)}{8}\right], \quad y \leq 45^\circ \text{ S}, \tag{4.85}$$

where τ^x is the zonal component of the wind stress. No meridional wind stress was imposed. The variable y denotes latitude. With this form of the imposed wind forcing, the wind-stress curl vanishes at 45° S.

In the first experiment, the model was integrated until the three-dimensional spatial average of the rate of change of temperature reached

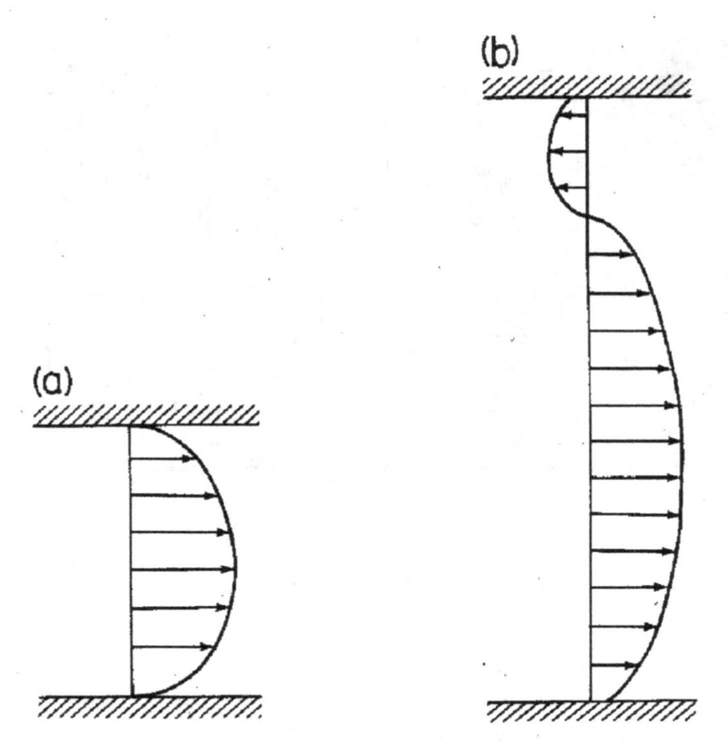

Fig. 4.19 The imposed values of the transport streamfunction in Matano's model. (a) Imposed streamfunction and velocity profile at the Drake Passage. (b) Imposed streamfunction and velocity profile at the open portion of the eastern boundary, between the Cape of Good Hope and the southern boundary of the model. The total transport is 120 Sv. Reproduced from Figure 7 of Matano (1993), with permission of the American Meteorological Society.

a value just under $10^{-8}\,°\mathrm{C\,s^{-1}}$. This occurred after three years of integration. The resulting transport streamfunction is shown in Figure 4.20.

It is interesting to note that Matano's reasonably well-resolved regional model, with a grid spacing of 0.5° near the coast, makes a systematic error in the position of separation of the Brazil current, while global-ocean general circulation models (e.g., Semtner and Chervin, 1988) produce the correct behavior of the Brazil current, regardless of resolution. This differs from experience in modeling the Gulf Stream, in which most numerical models up to that time had not been able to produce separation correctly at Cape Hatteras.

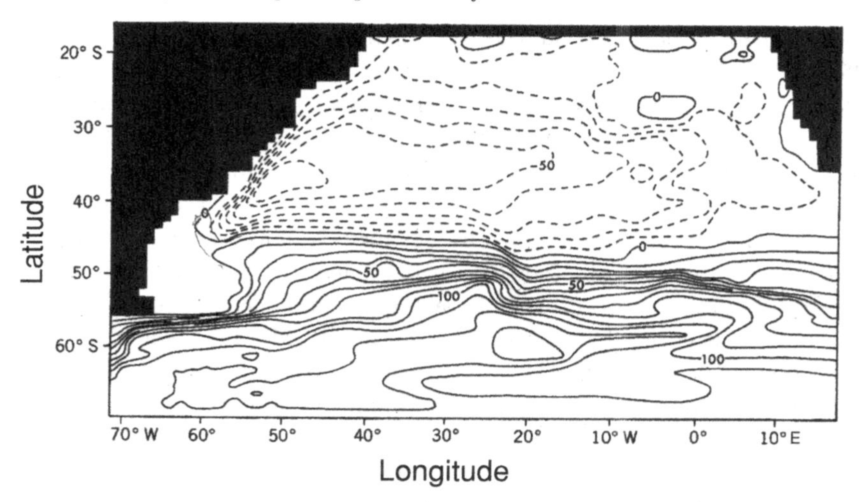

Fig. 4.20 Transport streamfunction after three years of integration of the primitive equation model. The contour interval is 10 Sv. Note the separation of the Brazil current at about 47° S, well south of the observed separation point. Reproduced from Figure 9 of Matano (1993), with permission of the American Meteorological Society.

In this simulation, the total transport in the western margins, including the Brazil current and the deep transport of North Atlantic Deep Water (NADW) is about 60 Sv, which is consistent with observations at the time. Less was known about the transport of the Malvinas current, but available data and earlier calculations showed estimates of about the same magnitude, while the model estimate of the Malvinas current was distinctly weak at 8 Sv. This is in line with estimates available to Matano. Most of these estimates were based on hydrography, with an assumed reference level. It is likely that the choice of reference level led to systematic underestimates of the transport in this largely barotropic current.

In the next series of experiments, Matano applied different forms of the wind forcing to the model. He was able to force the model to produce separation of the current near the correct latitude by imposing a form of the wind stress with the zero of the wind-stress curl at 40° S. This, however, is not consistent with available wind observations. This leaves the boundary conditions as candidates for the source of the problem.

In the first experiment with modified boundary conditions, the eastern boundary condition was modified so that the Agulhas current was turned off, i.e., there is no inflow at the eastern boundary. This made little difference.

Next, the effect of changing the western boundary condition was examined. One could concoct a simple linear theory based on the separation point being the point at which the transports of the Brazil current and the Malvinas current were equal and opposite. Given that the transport of the Brazil current is determined by Sverdrup balance, if the Malvinas current were also determined by Sverdrup balance, i.e., considered to be a western boundary current, then separation would occur at the zero of the wind-stress curl, which we know it does not. Matano then considered the hypothesis that the Malvinas current, rather than simply being a western boundary current, is a topographically trapped piece of the ACC. If this is the case, then the model's estimate of the transport of the Malvinas current could be increased by either increasing the transport of the ACC or increasing the proportion of the ACC that follows the f/H contours northward along the coast of Argentina.

Matano then performed two experiments to show that either of these possibilities could lead to northward displacement of the separation point. First, he increased the total flux of the ACC from 120 to 180 Sv, this latter value chosen to match the results from global ocean general circulation models. The reader should note that in the coarse resolution study of Cox (1989), described in Section 4.7, the transport of the ACC was greater still. In his last experiment, he kept the total transport to 120 Sv, but modified the transport profile at the Drake Passage as shown in Figure 4.21.

In the transport profile shown in Figure 4.21(b), the transport is concentrated in the northern part of the Drake Passage. A glance at Figure 4.17 shows that the f/H contours in that region continue northward along the coast of Argentina. In this case, if indeed the flow from the western inlet follows f/H contours, we would expect a greater proportion of the ACC water to contribute to the Malvinas current than was the case in the earlier experiment.

In both of these last experiments, the separation of the Brazil current was deflected northward to a point near the observed separation point. These experiments reveal a consistent picture of the separation of the Brazil current: the Brazil current is a western boundary current, resulting from western intensification according to the classical Sverdrup circulation theory as described by Stommel (1966). The Malvinas current, however, is only partly a direct result of the westward intensification of the Sverdrup equilibrium flow. It also includes a large contribution from

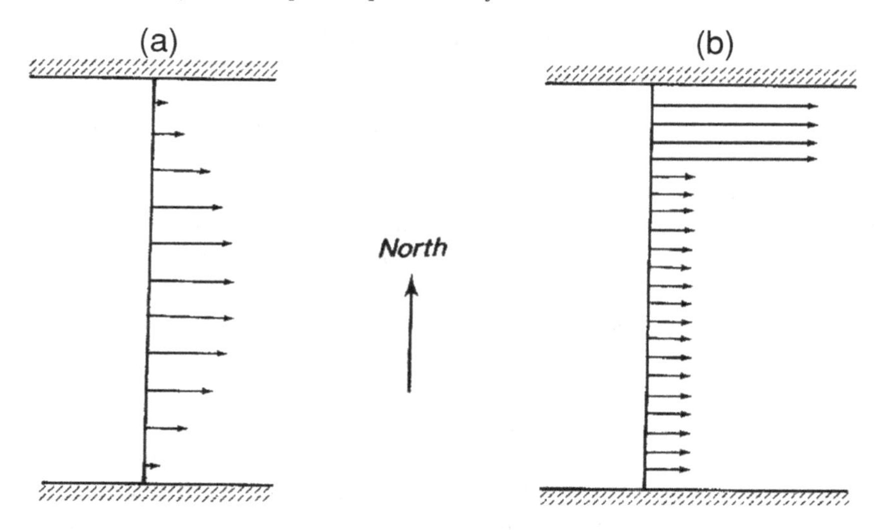

Fig. 4.21 Mass transport profiles imposed across the model Drake Passage: (a) as in the previous experiments; (b) as in the final experiment. Reproduced from Figure 14 of Matano (1993), with permission of the American Meteorological Society.

the ACC flowing through the Drake Passage and northward along the f/H contours following the coast of South America. Separation occurs at the point where the two currents are equal in magnitude. Because of the contribution from the ACC to the Malvinas current, that point is displaced northward from the zero of the wind-stress curl.

Since the publication of Matano's paper, new results have come to light that have led to a re-evaluation of the transports of the ACC, the Brazil current and the Malvinas current. Based on observations, some investigators believe the Malvinas current to be as strong as 70 Sv (see, e.g., Peterson *et al.*, 1996). Earlier estimates were based on geostrophic calculations that involved the assumption of a level of no motion which we now believe was too shallow to represent the predominantly barotropic structure of the Malvinas current.

So despite the fact that the details of Matano's work have been overtaken by subsequent events, it is valuable as a case study of what models can tell us. In particular, the location of the separation of the current is sensitive to the transport profile in the Drake Passage. Other influences, such as remote influence of the eastern boundary condition or slight changes in the wind stress that might be attributed to measurement error, could be ruled out by studies such as this one.

4.9 Generalized vertical coordinates

There are often advantages to replacing the vertical coordinate z with some other coordinate representation. In the numerical weather prediction community, pressure is considered to be the elementary vertical coordinate, and the use of z in atmospheric models is all but unknown. Arguments can be made in favor of pressure as a vertical coordinate in ocean models. Pressure, as opposed to depth, is the quantity directly measured by instruments. Moreover, pressure, rather than actual geometric depth, is the more dynamically relevant quantity. The pressure at a depth of 1000 m at the equator differs from the pressure at 1000 m in the polar oceans due to variation in apparent gravitational acceleration. This variation in acceleration is due to variation in centrifugal force and to the eccentricity of the shape of the earth. The effects are small but measurable, and may be relevant for models intended to describe oceanic motions on large spatial and temporal scales. The question of whether scientific insight is to be gained from modifying our models to account for this will be a matter of debate in coming years.

One can also use potential density as a vertical coordinate. This choice in its simplest form leads to the layer models, in which the ocean is considered as slabs of fluid of uniform density, with some basic assumption made about interfacial stress. These models are sufficiently widespread that they will be treated in a separate section.

We first consider models with vertical coordinates chosen so that the bottom is a coordinate surface. This choice of coordinates is very common in coastal models. The simplest and most common such coordinate choice is that of σ-coordinates. This is the common choice in numerical weather prediction, where the vertical coordinate is chosen to be $\sigma = (p - p_s)/p_s$, where p_s is the surface pressure.

If the bottom topography is given by $z = -D(x, y)$ and the mean surface is given by $z = 0$, we define the coordinate $\sigma = z/D < 0$, i.e., the range of σ is from -1 to zero, and the bottom is defined by the surface $\sigma = -1$. In this new coordinate system the bottom boundary condition takes the simple form

$$\frac{\mathrm{D}\sigma}{\mathrm{D}t}\Big|_{\sigma=-1} = 0, \tag{4.86}$$

where the total derivative on the left-hand side is evaluated at the bottom.

The hydrostatic relation can be derived from:

$$\frac{\partial p}{\partial z} = \frac{\partial p}{\partial \sigma} \times \frac{1}{D}, \tag{4.87}$$

so

$$\frac{\partial p}{\partial \sigma} = -\rho g D. \tag{4.88}$$

Calculation of the horizontal pressure gradient must account for the slope of the coordinate surfaces, so

$$\begin{aligned} \frac{\partial p}{\partial x}\Big|_z &= \frac{\partial p}{\partial x}\Big|_\sigma + \frac{\partial p}{\partial \sigma}\frac{\partial \sigma}{\partial x} \\ &= \frac{\partial p}{\partial x}\Big|_\sigma + \rho g \sigma \frac{\partial D}{\partial x}. \end{aligned} \tag{4.89}$$

The form of the pressure gradient term in σ-coordinates can lead to difficulties in the case of steeply sloping topography. In the case in which the density field depends on z alone, there is obviously no horizontal pressure gradient, but truncation error may prevent the two terms on the right-hand side of (4.89) from canceling in the discretized case. It is important to point out that the form of the expression dictates that truncation errors in both horizontal and σ directions contribute to these artificial pressure gradients. In general, refining the mesh in σ alone, while leaving the horizontal grid spacing Δx constant will not solve the problem. This is illustrated by a simple calculation.

Consider a simple two-dimensional difference scheme in σ-coordinates. Let $p_{i,j}$ be the pressure at $x = x_i$, $\sigma = \sigma_j$. Assume Δx and $\Delta\sigma$ are constant. Write the centered difference scheme:

$$p_x\Big|_\sigma \approx \frac{p_{i+1,j} - p_{i-1,j}}{2\Delta x}, \tag{4.90}$$

$$p_\sigma\Big|_x \approx \frac{p_{i,j+1} - p_{i,j-1}}{2\Delta \sigma}. \tag{4.91}$$

Now assume the bottom is given by $z = -D(x)$ and $\sigma = z/D$. We then have

$$p_x\Big|_z \approx \frac{p_{i+1,j} - p_{i-1,j}}{2\Delta x} - \frac{\sigma D_x}{D}\left(\frac{p_{i,j+1} - p_{i,j-1}}{2\Delta \sigma}\right). \tag{4.92}$$

Now assume the bottom slope is constant, i.e., $D(x) = H_0 - sx$. For the purpose of this simple calculation, we assume that we know the density profile $\rho(z)$ exactly. We then have

$$\frac{p_{i+1,j}-p_{i-1,j}}{2\Delta x} = \frac{-g}{2\Delta x}\int_{\sigma_j[H_0-s(x_i-\Delta x)]}^{\sigma_j[H_0-s(x_i+\Delta x)]}\rho\,\mathrm{d}z, \tag{4.93}$$

$$\frac{p_{i,j+1}-p_{i,j-1}}{2\Delta\sigma} = \frac{-g}{2\Delta\sigma}\int_{\sigma_{j-1}(H_0-sx_i)}^{\sigma_{j+1}(H_0-sx_i)}\rho\,\mathrm{d}z. \tag{4.94}$$

Now suppose we have an exponential density profile, i.e., $\rho = \delta\exp(z/D_\rho)$ where D_ρ is the scale of the density variation. Evaluation of the integrals in (4.93) and (4.94) leads to:

$$\begin{aligned} p_x\big|_\sigma - \frac{\sigma D_x}{D}p_\sigma = p_x\big|_\sigma + \frac{\sigma s}{D}p_\sigma \\ &\approx gD_\rho\delta\sigma_j\mathrm{e}^{\sigma_j D/D_\rho} \\ &\times\left[\frac{\sinh(\sigma_j s\Delta x/D_\rho)}{\sigma_j\Delta x} - s\frac{\sinh(\Delta\sigma D/D_\rho)}{\Delta\sigma D}\right]. \end{aligned} \tag{4.95}$$

From this expression, we can see explicitly that the pressure gradient vanishes in the limit as $\Delta\sigma$ and Δx decrease to zero, but for fixed Δx, the pressure gradient does not approach zero as $\Delta\sigma \to 0$. In fact, for small values of Δx and $\Delta\sigma$, the expression in brackets on the right-hand side of (4.95) is well approximated by

$$\frac{1}{6}\frac{sD^2}{D_\rho^3}\left[\frac{\sigma_j^2 s^2\Delta x^2}{D^2} - \Delta\sigma^2\right]. \tag{4.96}$$

This expression vanishes when

$$\Delta\sigma^2 = \frac{\sigma_j^2 s^2\Delta x^2}{D^2}, \tag{4.97}$$

leading to the result that the error actually increases as $\Delta\sigma$ decreases beyond the value on the right-hand side of (4.97). We therefore obtain the expected behavior of the error decreasing as $\Delta\sigma$ decreases as long as

$$\frac{\sigma_j^2 s^2\Delta x^2}{D^2} \le \Delta\sigma^2,$$

or, more generally,

$$\left|\frac{\sigma_j D_x}{D}\right|\Delta x \le \Delta\sigma. \tag{4.98}$$

The condition (4.98) is often referred to as the "hydrostatic consistency" condition; see e.g., Haney (1991). This is interpreted geometrically in terms of the two terms on the right-hand side of (4.92). If the topography slopes sharply and $\Delta\sigma$ is small, then (4.93) is calculated over a smaller range of the water column than (4.94), and therefore (4.93). Examination of (4.93) and (4.94) reveals that (4.98) is exactly the condition that the ranges of the two integrals overlap.

Other examples of this type can be found in the literature. Haney (1991) performed a series of calculations similar to the above, intended to illustrate the problems associated with the form of the pressure gradient in σ-coordinates. He used the trapezoid rule to approximate the integral of the buoyancy from a given level to the surface in order to calculate the pressure at each gridpoint. With that scheme, for equally spaced σ-levels, the pressure gradient vanishes identically when the density depends linearly on depth and is independent of x and y (see Exercise 4.3). Unequal spacing of the σ-levels can lead to spurious pressure gradients in the discrete system. Haney also performed a series of calculations with density profiles given by the baroclinic modes of the North Pacific, as calculated from density profiles. He illustrated in that case the behavior observed in our simple calculations: for fixed horizontal grid spacing, the error can actually increase as the spacing of the σ-levels decreases.

Yet another similar calculation was performed by Beckmann and Haidvogel (1993). In their calculations, they assumed the derivatives with respect to σ to be evaluated exactly, and calculated the pressure gradient in a straightforward difference scheme for polynomial density profiles. As in Exercise 4.3 they found that the pressure gradient vanishes for horizontally uniform density distribution, in which the density depended linearly on z. A spurious gradient appeared in case of quadratic and higher degree dependence of the density upon depth.

In further calculations, Beckmann and Haidvogel (1993) applied a σ-coordinate model to the problem of an isolated seamount in an f-plane channel. The geometry of the calculation was intended to simulate the situation at Fieberling Guyot, a tall, narrow seamount in the subtropical eastern North Pacific Ocean. Their model geometry is shown in Figure 4.22. The seamount is assumed to have Gaussian shape with standard deviation equal to 25 km, and maximum height 4050 m above a flat bottom, which approaches a depth of 4500 m far from the seamount.

The model uses centered differences on an Arakawa C-grid. A rigid lid is imposed at the surface. Vertical variability of the model quantities is represented by a series of Chebyshev polynomials. Nonlinear terms were omitted from the dynamical equations, and no explicit diffusion in the σ-direction was added. Pressure due to the resting stratification was

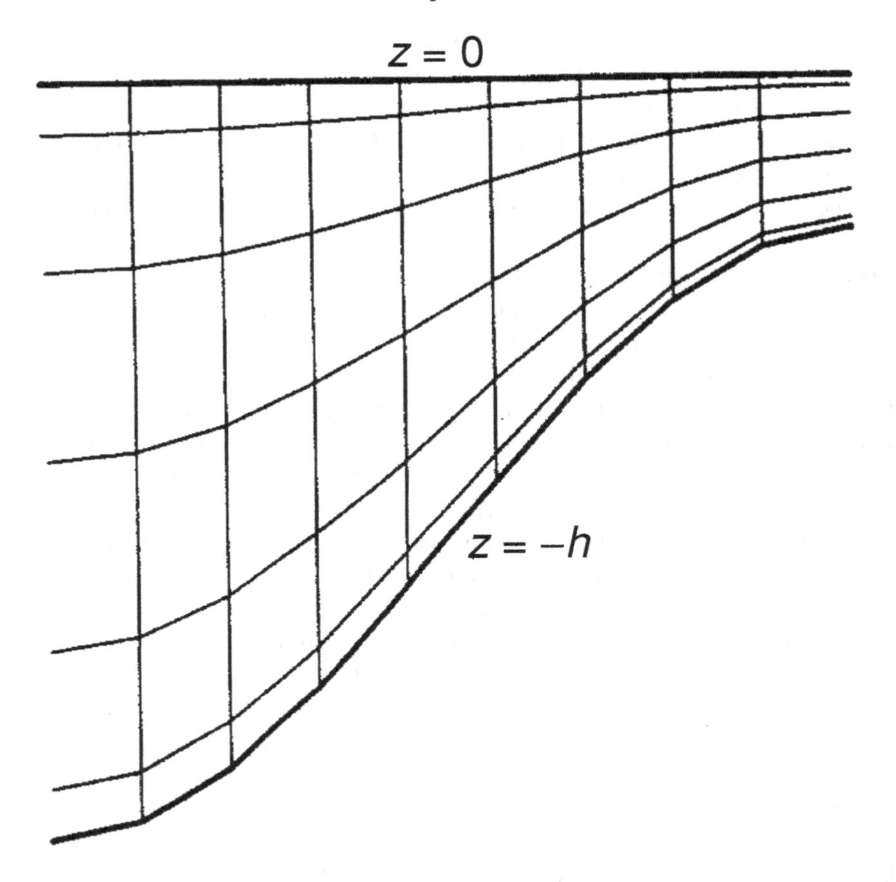

Fig. 4.22 Cross-section of seamount geometry. The stretched horizontal grid is most clearly evident at the boundaries. Reproduced from Figure 2 of Beckmann and Haidvogel (1993), with permission of the American Meteorological Society.

removed, so pressure was calculated as a deviation from the hydrostatic pressure resulting from some mean stratification. Perturbation densities that varied linearly and exponentially with depth were considered. Computations were performed with no external forcing, and with periodic forcing intended to simulate weak diurnal tidal forcing. In general, errors were greater in the unforced cases, shown here. The reader should refer to the original article for the forced results.

Calculations were performed with linear and exponential perturbation density profiles. The linear profiles gave rise to solutions that were

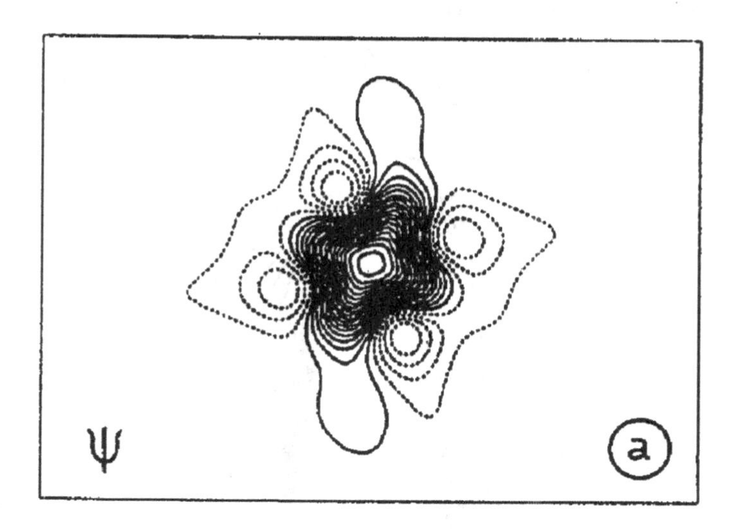

Fig. 4.23 External streamfunction after 100 days of computation for the sigma coordinate model with exponential stratification. The contour interval is 0.05 Sv. Reproduced from Figure 7a of Beckmann and Haidvogel (1993), with permission of the American Meteorological Society.

correct up to machine accuracy. In the exponential case, a perturbation of the form $-0.1\,\mathrm{kg\,m^{-3}} \times \exp(z/1000\,\mathrm{m})$ was added to a resting stratification of the form $1028\,\mathrm{kg\,m^{-3}} - \Delta_{z\rho} \times \exp(z/1000\,\mathrm{m})$, where $\Delta_{z\rho}$ is the difference between the density at the surface and the density of the bottom. Results were expressed in terms of the resolution of the variation of the topography $r = (h_i - h_{i-1})/(h_i + h_{i-1})$, where h_i is depth of the water at the ith gridpoint, and the Burger number $S = N_0 H_0/(f_0 L)$ where $L = 25\,\mathrm{km}$ is the scale of the variation of the topography, $H_0 = 4500\,\mathrm{m}$ is the total depth and N_0 is the buoyancy frequency. We may identify the parameter r as being approximately one-half the quantity $D_x \Delta x/D$ that appears in our expression (4.98) for hydrostatic consistency in the earlier example.

Figure 4.23 shows the external streamfunction following 100 days of simulation for the unforced case with exponential stratification. This run was conducted with $S = 1.5$ and r taking a maximum value of 0.21. The maximum amplitude of the external streamfunction is in excess of 0.8 Sv in a stationary eightfold symmetric pattern.

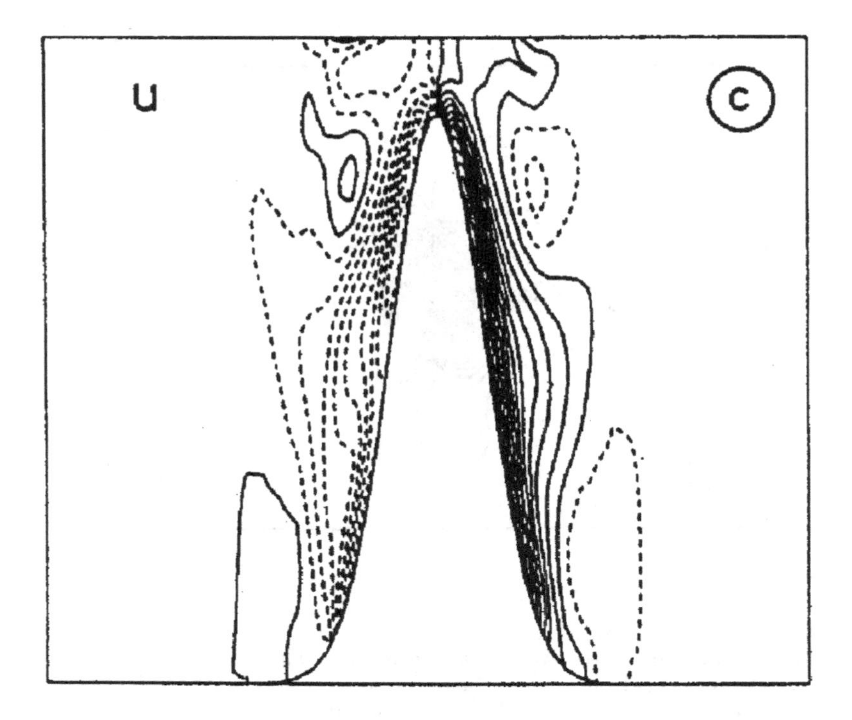

Fig. 4.24 Section through the center of the seamount, facing upchannel, of the along-channel component of the velocity after 100 days of integration. The contour interval is $2.5 \times 10^{-3}\,\mathrm{m\,s^{-1}}$. Reproduced from Figure 7c of Beckmann and Haidvogel (1993), with permission of the American Meteorological Society.

The along-channel velocity component after 100 days of integration is shown in Figure 4.24. The near antisymmetry of the contours indicates a circulation around the seamount, with maximum speed of about $3.6 \times 10^{-2}\,\mathrm{m\,s^{-1}}$.

Beckmann and Haidvogel went on to examine the dependence of the velocity in the isopycnal unforced case upon the parameters r and S by performing a series of 10-day runs, with S ranging from 0.0 to 6.0 and r taking the values of 0.173, 0.211, 0.266 and 0.310, corresponding to different choices of grid spacing. As expected, the velocity increases with increasing S and increasing r. The computation actually failed before day 10 of model time for the strongest stratification and coarsest resolution (i.e., highest r). Most of the velocities are of the order of $1.0 \times 10^{-2}\,\mathrm{m\,s^{-1}}$ after 10 days. Given the 100-day run described above, we conclude that the fields are not fully developed. Much higher values shown in the table of runs for the largest values of S and r in the study probably represent incipient instability. Beckmann and Haidvogel go on to investigate different numerical schemes, whose performance differs in detail from their simplest scheme, but follow the same general pattern.

Haidvogel and Beckmann (1999) suggest $r \leq 0.2$ as a rule of thumb for ensuring stability of sigma coordinate calculations. If we recast (4.98) as $r \leq 2\Delta\sigma/\sigma_j$, we can see that this choice will result in violation of hydrostatic consistency unless r is chosen $<< 0.2$ or the resolution in σ is very coarse.

The expression in brackets in (4.96) is quadratic, reflecting the second-order accuracy of the discretized calculation. One might therefore hope that higher-order methods might lead to smaller errors in the pressure gradient due to the balance of terms in (4.89). McCalpin (1994) performed a series of calculations similar to those performed by Beckmann and Haidvogel, but with comparisons of second- and fourth-order discretizations in the horizontal directions. Like Beckmann and Haidvogel, McCalpin's calculations were performed in a periodic channel with a seamount with Gaussian profile, and vertical variability was represented by a series of Chebyshev polynomials. The horizontal resolution was varied by varying the horizontal scale of the seamount. All model runs showed patterns resembling those found by Beckmann and Haidvogel. Comparison of results of calculations by different methods is shown in Figure 4.25, in which the erroneous pressure gradient arising at the initial time from imbalance of terms in (4.89) are depicted in terms of the maxima of the geostrophically balanced velocities. The resolution increases to the right on the x-axis and so the maximum velocities decrease to the right as expected. For all but the coarsest resolutions, the fourth-order method does better than the second-order method, for both the uniform and stretched grids. Note that the scale of this figure is logarithmic, so the slopes of the curves reflect the orders of accuracy of the methods employed. Careful examination of the slopes of the curves for the uniform grid cases show values very close to those one would expect from the orders of convergence.

It is possible to avoid some of the problems with the σ-coordinate model, while retaining some of its advantages over the z-coordinate system by implementing a general coordinate transformation. Gerdes (1993) implemented such a general coordinate system with the Bryan–Cox model. Gerdes refers to his generalized vertical coordinate as an

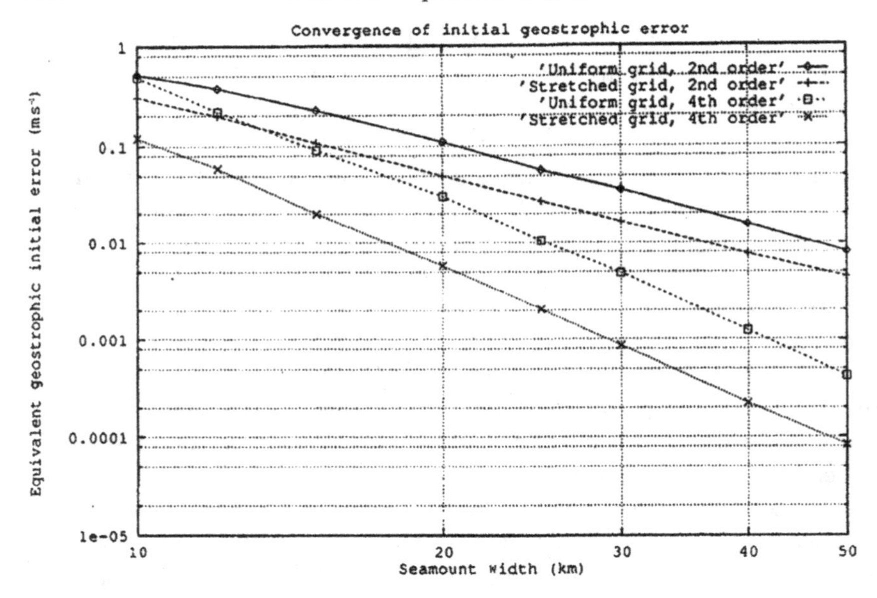

Fig. 4.25 Maximum value of geostrophic velocity required to balance the initial pressure gradient error. Reproduced from Figure 3 of McCalpin (1994), with permission of John Wiley and Sons, Ltd.

"s-coordinate"; the layout of the s-coordinate model is shown schematically in Figure 4.26.

The s-coordinate model can represent the topography more faithfully than a z-coordinate model with the same number of levels in the vertical. This is illustrated by an example of a barotropic channel on the f-plane with exponential bottom topography. Since an analytical form of the dispersion relation for this problem is available, the properties of s- and z-coordinate systems with similar numbers of levels can be compared. The numerical dispersion relations, along with the analytical form of the dispersion relation, are shown in Figure 4.27.

The numerical experiment consisted of forcing the system with cross-channel wind stress with normalized frequency $\nu = 1/7$ and sinusoidal cross-channel structure with a single full wave. The forcing was chosen to have a white spectrum in the along-channel direction, so the forcing contained all resolved wavenumbers. Figure 4.28 shows a summary of the results of this experiment.

A spectral analysis in along-channel wavenumber space at every latitude was performed on the streamfunction Ψ after a fairly long

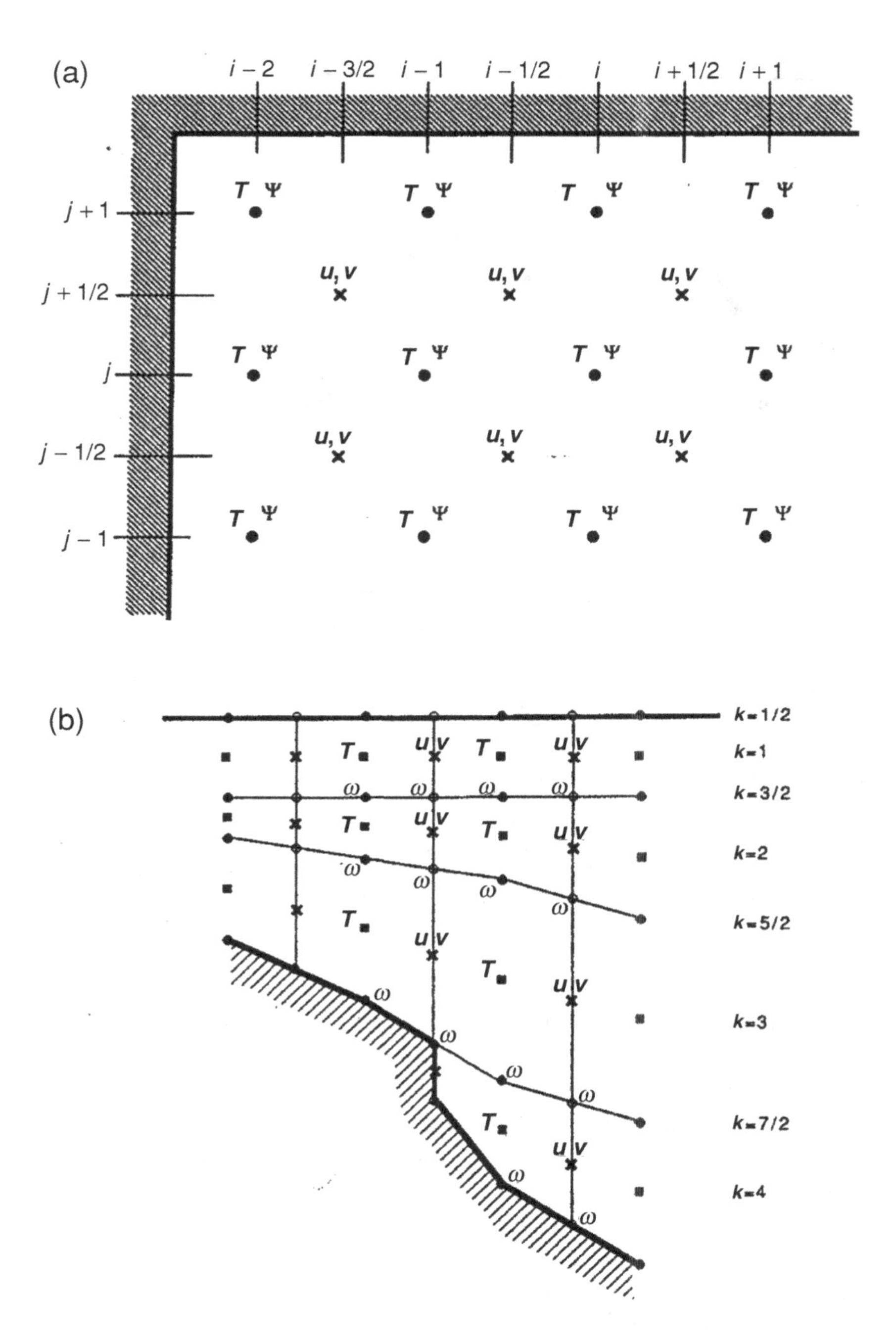

Fig. 4.26 (a) Horizontal and (b) vertical arrangement of variables in the generalized vertical coordinate model. T marks tracer gridpoints, u, v indicate horizontal velocity components, Ψ indicates the location of external-mode streamfunction gridpoints and ω is the transformed vertical velocity. The bottom $z = -H(x, y)$ was chosen not to coincide with a coordinate surface in the zonal section (b) in order to indicate how intersections of coordinate surfaces and topography are handled. Reproduced from Figure 1 of Gerdes (1993), with permission of the American Geophysical Union.

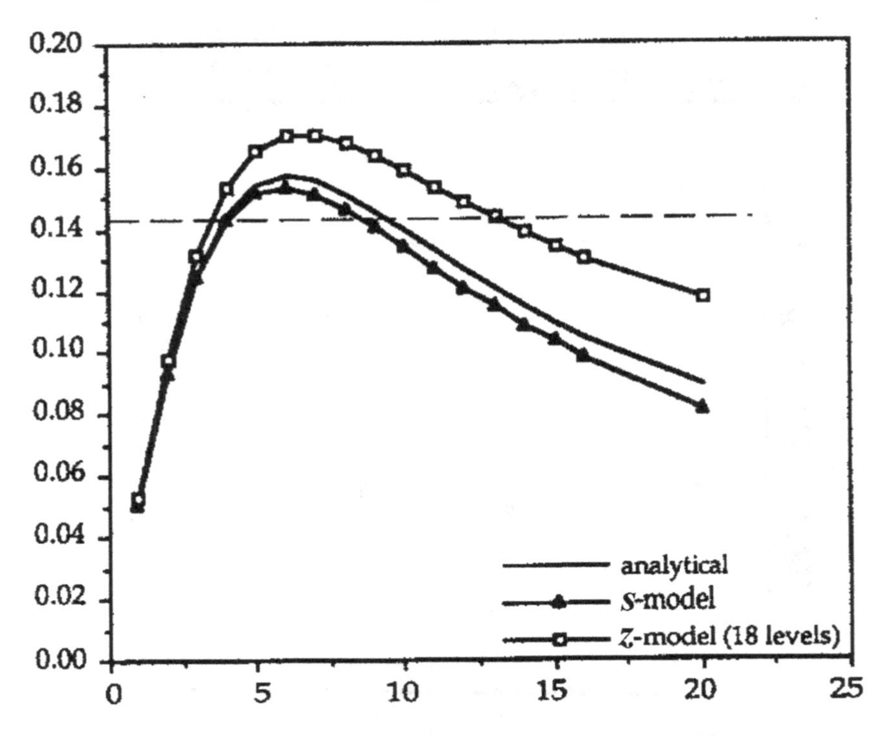

Fig. 4.27 Dispersion relation for the homogeneous, exponential bottom topography experiment. The analytical case and the result of the eigenvalue analysis for the s-coordinate model and the z-coordinate model with 18 levels in the vertical are shown. The dashed line represents the normalized frequency $\nu = 1/7$ of the forcing applied in the numerical experiments. Reproduced from Figure 2 of Gerdes (1993), with permission of the American Geophysical Union.

integration of the model, with the results one would expect from examination of Figure 4.27. The spectra of both models were strongly peaked at the wavenumbers where the curves corresponding to the models crossed the dashed line in Figure 4.27. Both models showed peaks near wavenumber 4, but while the s-coordinate model showed a peak near wavenumber 9, which is close to the value expected from the analytic solution, the z-coordinate model showed a peak at a wavenumber which was much too high. The s-coordinate model is therefore more efficient in this application. Similar fidelity to the exact solution can only be had in a z-coordinate model with more levels, and therefore greater expense.

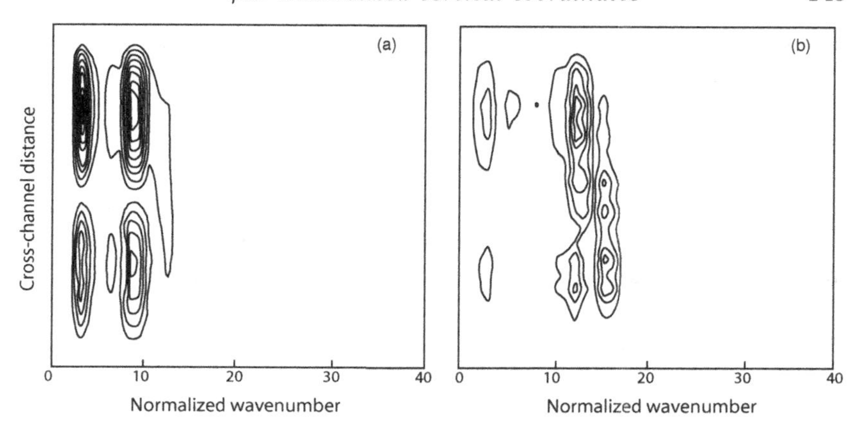

Fig. 4.28 Spectrum of the streamfunction Ψ as a function of zonal wavenumber and latitude. The deeper part of the channel is at the top of the figure. (a) s-coordinate model; (b) z-coordinate model. Redrawn from Figure 4 of Gerdes (1993), with permission of the American Geophysical Union.

Details of construction of specific schemes can be found in the literature. Many of the schemes in common use were derived by the methods described by Arakawa and Suarez (1983). They investigated a family of schemes in which the pressure gradient force generates no circulation of vertically integrated momentum along contours of bottom topography, and in which the finite-difference analogs of the energy conversion term have the same form in the kinetic energy and the thermodynamic equations. Without these constraints, truncation errors can lead to spurious sources or sinks of total energy or vertically integrated vorticity.

The consequences of these pressure gradient artifacts for practical models can be subtle. Mellor *et al.* (1997) claim that the hydrostatic consistency criterion places unrealistic demands on practical computational models, and show an apparently satisfactory calculation of an idealized problem with steep topography in which the pressure gradient artifacts disappear. At first, spurious velocities arise through a geostrophic adjustment process. These velocities then disappear on frictional timescales, leaving behind spurious density gradients that just balance the terms in the expression for the pressure gradient. The significance of these erroneous density gradients is a matter of interpretation. No complete dynamical description of the adjustment process has yet been given. The geostrophic adjustment process alone cannot explain the details of the evolution of the velocity field, even in an idealized simulation; see, e.g., McCalpin (1994).

4.10 Layer models

In this section we follow the approach of Bleck and Smith (1990). Conceptually, in formulating layer models in their purest form, we view the ocean as a stack of slabs of immiscible fluids with different densities. In this simple picture, the interfaces between the slabs are material surfaces. The layer model is a common schematic view of the ocean.

In numerical implementation, sharp density gradients can be represented more faithfully than they can in Cartesian coordinate models. More importantly, they conform readily to the widely accepted view that mixing takes place along isopycnal surfaces, and therefore constitute a more efficient representation of diffusion processes.

One problem with the theorist's simple picture of the ocean as a layered medium arises due to the complex equation of state of seawater. In fact, the use of potential density referred to some fixed pressure as the vertical coordinate in a layer model will result in an unrealistic representation of the ocean. One can find cases in which a parcel of seawater, call it parcel "A," has greater surface-referenced potential density than another parcel "B," while at depths of a few thousand meters, "A" has greater buoyancy. Clear evidence of this effect can be seen in maps of surface-referenced potential density in the North Atlantic, in which large potential density inversions occur at depth.

It has been suggested that reference pressures other than the surface be used, and maps of potential density in the North Atlantic relative to 2000 m do not show density inversions, but those potential density values do not yield reliable results in calculations of thermal wind shear. An example presented by Sun *et al.* (1999) shows that no single choice of reference level will allow an isopycnic model to represent the pressure gradients that are thought to drive the subsurface circulation in the Atlantic, which is characterized by northward-flowing upper Antarctic Intermediate Water nearest the surface, lying just above southward-flowing North Atlantic Deep Water, lying in turn above northward-flowing Antarctic Bottom Water.

The question of how to formulate a realistic layer model is the subject of controversy at this writing. Two different views are expressed by de Szoeke (1998) and McDougall (1987). While we expect this to be the subject of intense debate in the ocean modeling community in the coming years, we also expect to see continued extensive use of layer models. We therefore present the basic layer model formulation.

In the simplest layer model formulation, the layer boundaries are material surfaces, i.e., they move with the flow and no fluid crosses them. The pressure gradient on coordinate surfaces is derived as a simple change of coordinates. Let (x_1, z_1) and (x_2, z_2) be points on the interface between two layers, and let the pressures at these two points be p_1 and p_2 respectively. Assume the interface between the layers is given by $\zeta(x, y) =$ constant. Let p_0 be the pressure at the point (x_1, z_2). We have

$$p_2 - p_0 = (p_2 - p_1) + (p_1 - p_0),$$

so

$$p_x\big|_z \delta x = p_x\big|_\zeta \delta x - p_z \delta z,$$

where $\delta x = x_2 - x_1$ and $\delta z = z_2 - z_1$. Dividing by δx and taking the limit as $\delta x \to 0$ leads to

$$\begin{aligned} p_x\big|_z &= p_x\big|_\zeta - p_z z_x\big|_\zeta \\ &= p_x\big|_\zeta + \rho g z_x\big|_\zeta, \end{aligned}$$

where the hydrostatic relation was applied to derive the second of the above equations. We now define the specific volume $\alpha \equiv 1/\rho$, so the pressure term in the momentum equation becomes

$$\frac{1}{\rho} p_x\big|_z \equiv \alpha p_x\big|_z = (\alpha p + gz)_x\big|_\zeta, \tag{4.99}$$

since α is constant within layers. The quantity $M = \alpha p + gz$ is known as the "Montgomery Streamfunction."

We can now write the equations of motion for the layer model as Bleck and Smith write them. The pressure thickness of the mth layer is $\Delta p_m = g\rho_m(z_m - z_{m+1})$. In terms of the variables as we have described them here, the equations of motion are

$$\frac{\partial u}{\partial t} + \frac{\partial}{\partial x}\frac{\mathbf{u}\cdot\mathbf{u}}{2} - (\zeta + f)v = -\frac{\partial M}{\partial x} + \frac{1}{\Delta p}(g\Delta\tau + \nabla\cdot\nu\Delta p\nabla u), \tag{4.100}$$

$$\frac{\partial v}{\partial t} + \frac{\partial}{\partial y}\frac{\mathbf{u}\cdot\mathbf{u}}{2} + (\zeta + f)u = -\frac{\partial M}{\partial y} + \frac{1}{\Delta p}(g\Delta\tau + \nabla\cdot\nu\Delta p\nabla v), \tag{4.101}$$

$$\frac{\mathrm{D}\alpha}{\mathrm{D}t} = 0, \tag{4.102}$$

where $\Delta\tau$ is the difference in stress between the top and bottom of the layer, ν is the kinematic viscosity and $\mathbf{u} = (u, v)$ is the velocity vector, as is customary.

We may compare this general formulation to the simple two-layer model introduced in (3.5)–(3.8). In (3.5), the momentum equation in the upper layer, the pressure term is given by $g(h_1 + h_2)_x$. At the surface, variations in atmospheric pressure are neglected, so the Montgomery streamfunction is simply gz, which, at the surface, becomes $g(h_1 + h_2)$. In the second layer, at a point just below the interface, $\alpha = 1/\rho_2$ and the pressure is $\rho_1 g h_1$ and $z = h_2$, so the Montgomery streamfunction is given by

$$M \equiv \alpha p + gz = \frac{\rho_1}{\rho_2} g h_1 + g h_2,$$

and the pressure terms in (3.5) and (3.7) are exactly the gradients of the Montgomery streamfunction.

Serious numerical difficulties arise when the thickness of one or more layers vanishes. This can happen when a layer outcrops at the surface, or when the interface between two layers intersects a sloping boundary. In this situation, the layer depth will, in general, be continuous as a function of x and y, but not differentiable. Since highly accurate numerical methods represent high wavenumber Fourier components more faithfully than less accurate ones, we expect to observe overshoot, and therefore negative layer depth, in integrations with accurate difference methods. At first glance, one might not expect this to be a major problem, since the overshoot is a manifestation of truncation error and should therefore be small, but a formally negative layer depth leads to a formally imaginary wave speed, which changes the character of the equations of motion. If one imagines the simplest relevant case, i.e., the shallow-water equations, a negative layer depth changes the local character of the equations from hyperbolic to elliptic. We therefore expect this situation to be catastrophic for the calculation, since the initial value problem for an elliptic equation is known to be ill-posed.

Outcropping will, in fact, occur in basin scale models, as illustrated by the simple calculation of Bogue *et al.* (1986). Write the linearized reduced gravity equations with a simple linear drag law:

$$-fDv = -DD_x + \lambda\tau^{(x)} - \epsilon u, \tag{4.103}$$

$$fDu = -DD_y + \lambda\tau^{(y)} - \epsilon v, \tag{4.104}$$

$$\nabla \cdot (Du) = 0, \tag{4.105}$$

where D is the depth of the active layer, the velocity u has components

u and v, $\tau^{(x)}$ and $\tau^{(y)}$ are the stress components. The stress scale λ is given by

$$\lambda = \frac{LW}{g'\rho_0 d^2},$$

where L is the basin scale, W is the stress scale and d is the resting depth of the upper layer.

Now consider (4.103). Assume a schematic form of the wind stress with $\tau^{(y)} = 0$ and $\tau^{(x)}$ depending sinusoidally on the meridional coordinate y. Friction is only important near the western boundary, and therefore contributes significantly to (4.104) rather than (4.103). Equation (4.103) then reduces to

$$-fvD = DD_x - \lambda \cos \pi y. \tag{4.106}$$

We may integrate this equation to find

$$-2f \int_x^1 Dv \, \mathrm{d}x = -[h_{\mathrm{E}}^2 - D^2(x)] - 2\lambda(1-x)\cos \pi y, \tag{4.107}$$

where h_{E} is the layer depth at the eastern boundary. At the eastern boundary the depth must be independent of y, or there would be a geostrophically balanced flow into the solid boundary according to (4.104). Geostrophic balance does not hold at the western boundary where friction contributes significantly.

The integral must vanish at $x = 0$, so we must have

$$h_{\mathrm{E}}^2 - D_{\mathrm{W}}^2(y) = -2\lambda \cos \pi y, \tag{4.108}$$

where D_{W} is the depth of the active layer at the western boundary. When $0 < y < 0.5$, we must have $D_{\mathrm{W}} > h_{\mathrm{E}}$, but when $0.5 < y < 1$, we must have $D_{\mathrm{W}} < h_{\mathrm{E}}$, so for λ sufficiently large, there will be some latitude y_c where $h_{\mathrm{E}}^2 = -2\lambda \cos \pi y_c$. At this latitude, the active layer must outcrop. North of y_c, the boundary of the active region must separate from the western boundary.

In order to avoid the problems associated with vanishing layer depths, we would like our schemes to preserve monotonicity. This will ensure that the problems associated with overshoot will not occur. If we write a general explicit numerical scheme as

$$u_i^{n+1} = Qu_i^n \equiv H(u_{i-q}^n, \ldots, u_i^n, \ldots, u_{i+q}^n), \tag{4.109}$$

it is said to be *monotone* if

$$\frac{\partial H}{\partial u_{i-j}^n} \geq 0$$

for $|j| \leq q$. This means that the effect of upward displacement of one of the gridded solution values results in upward displacement of the others. Clearly the opposite case, in which $\partial H/\partial u_{i-j}^n \leq 0$, is the case in which steepening of a front can result in oscillations in space. Monotone methods can be shown to be monotonicity preserving, i.e., if the grid function $\mathbf{u}$ is monotone, then $Q\mathbf{u}$ is monotone; see, e.g., Sod (1985) or Leveque (1992). Unfortunately, it can also be shown that monotone methods in which the function (4.109) is linear in its arguments can be no more than first-order accurate. This accuracy limitation is, in general, unacceptable, so specialized methods have been developed to deal with these problems. These specialized methods usually involve so-called "flux limiters," which are highly nonlinear functions of the fluxes of mass and momentum; the formulas governing these methods are assumed to be in conservation form, i.e., the scheme should mimic the general conservation forms (3.24) and (3.25).

The widely used MICOM model (Bleck and Smith, 1990) uses a version of the flux-corrected transport method, which is only first order in the neighborhood of steep gradients. This and other methods for dealing with the case in which the layer depths go to zero are described in Sod (1985) or Leveque (1992).

The results of a comparison among five different models was reported in Chassignet *et al.* (2000). The models were the Modular Ocean Model (MOM), an advanced version of the Bryan–Cox code, DieCAST (see, e.g., Dietrich, 1997), another z-coordinate model, the Princeton Ocean Model (POM; Blumberg and Mellor, 1987), a σ-coordinate model, the Regional Ocean Modeling System (ROMS), an s-coordinate model that evolved from the Semi-Spectral Primitive Equation Model (SPEM) and the s-coordinate Rutgers University Model (SCRUM; see Song and Haidvogel, 1994), and MICOM. Haidvogel and Beckmann (1999) contains compact descriptions of MOM, SPEM, MICOM and SEOM. All models were used to perform a simulation of the North Atlantic with 10–20 layers or levels in the vertical, and barely eddy-permitting horizontal resolution of 0.5° for most, with 0.75° resolution being used in the ROMS implementation. A similar comparison was described by Willebrand *et al.* (2001). The domain was bounded at the south and north at 6° N and 50° N respectively. The model state was restored to climatology through 3° sponge layers at the northern and southern boundaries. In most of the simulations another sponge layer was added in the Gulf of Cadiz to represent the Mediterranean. The MOM simulation actually included part of the Mediterranean. There were other fairly minor differences among the model implementations.

Atmospheric forcing fields were derived from the Comprehensive Ocean-Atmosphere Data Set (da Silva *et al.*, 1994). Initial conditions were given by the Levitus (Levitus, 1982) climatology for January. The models were initially spun up for at least ten years of model time, which is sufficient for basic local adjustment, but too short to produce a fully developed thermohaline circulation. An interval of 10 to 20 years is generally considered the time required for the circulation to adjust to the wind forcing. This scale is set by propagation of higher-mode baroclinic Rossby waves. We saw an example of the beginning of this adjustment process in Figure 4.3. The adjusted state reached by models following this process is sometimes referred to as "dynamic quasi-equilibrium," (see, e.g., Döscher *et al.*, 1994). The authors report (Chassignet *et al.*, 2000) that the velocity fields reached this state after five years. The last three years of each run were used in the analysis.

The meridional overturning streamfunctions for the three-year averaged flows are shown in Figure 4.29.

All of the models exhibit the expected thermohaline circulation, with maximum transport between 1000 and 1500 m. The maximum transport varies surprisingly among the models, from a low of 12 Sv in ROMS to 21 Sv for MOM.

The level models MOM and DieCAST exhibit spurious upwelling near 35° N, as do ROMS and POM, which are similar to level models near the surface. This is a consequence of the excessive cross-isopycnal mixing characteristic of level models, and is absent from the MICOM result.

A curious pattern of secondary overturning cells centered around 1000 m appears in POM. A similar pattern occurs in the corresponding figure in the comparison reported in Willebrand *et al.* (2001), and does not appear in any of the other simulations. The σ-coordinate model used in the comparisons reported in Willebrand *et al.* (2001) was based on SPEM (Haidvogel *et al.*, 1991), so this is not related to particular features of POM. Given this, it is curious that ROMS does not exhibit this pattern.

MOM, DieCAST and ROMS show a deep overturning cell at the south, i.e., the lower right corner of the basin. This is associated with Antarctic Bottom Water, and is driven by large-scale pressure gradients that cannot be represented by MICOM, or any other model that uses potential density as a coordinate, as noted earlier in this section.

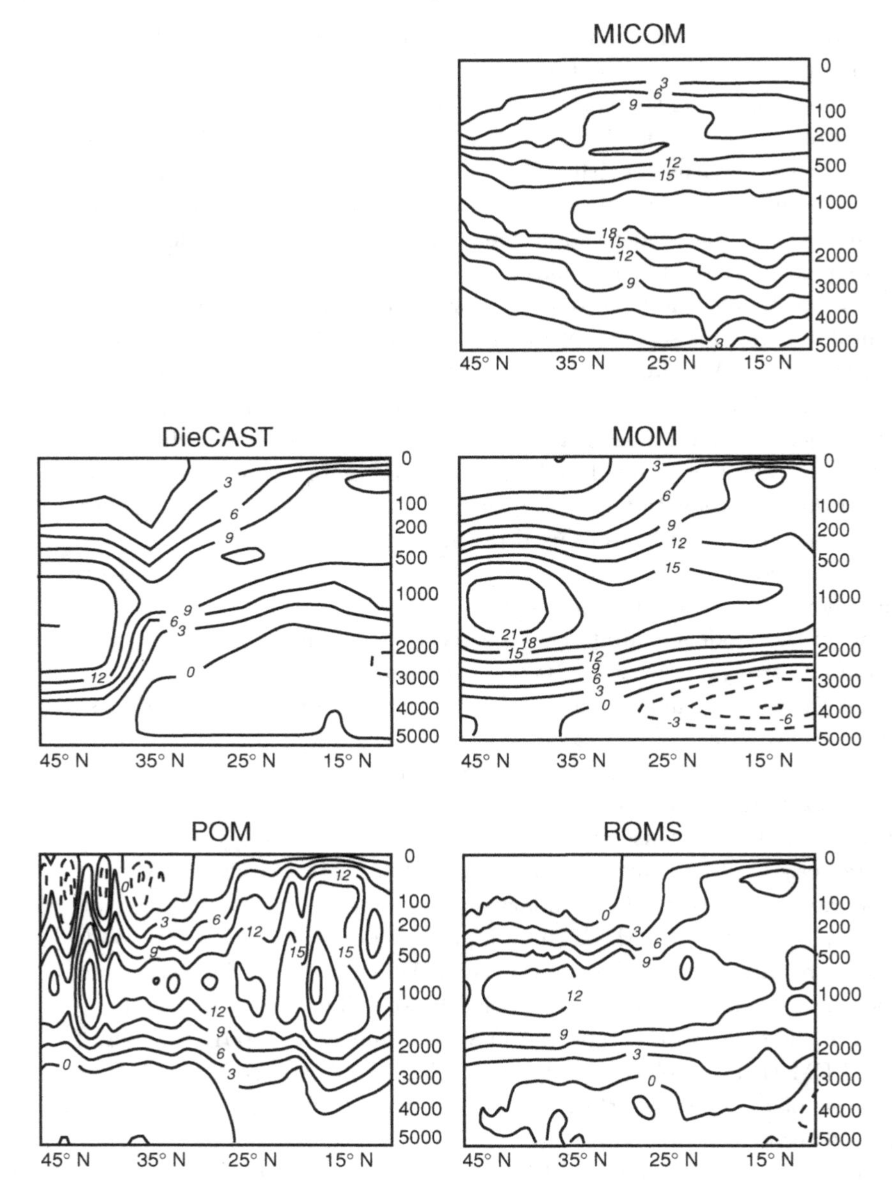

Fig. 4.29 Meridional overturning streamfunction for the five models. The contour interval is 3 Sv. Redrawn from Figure 2 of Chassignet *et al.* (2000), with permission of Elsevier.

Talley *et al.* (2003) presented a purely data-based estimate of the meridional overturning streamfunctions in the Atlantic, Pacific and Indian Ocean basins. Those calculations were based on geostrophic velocities derived from hydrographic sections along with estimated Ekman transports from the National Centers for Environmental Prediction (NCEP) re-analysis. Similar calculations performed with the Hellerman and Rosenstein (1983) winds produced transports that differed by no more than 1 Sv. The analysis of Talley *et al.* (2003) shows a structure similar to those depicted in Figure 4.29. The overturning transport takes its maximum of just over 18 Sv at a depth of about 1500 m. The deep overturning cell associated with the Antarctic Bottom Water has a maximum transport of about 8.5 Sv. Talley *et al.* (2003) estimate the errors in their calculations to be about 3–5 Sv, so all the model results are in the right range.

Upon reflection it seems that the evident similarity of the model results (Chassignet *et al.*, 2000; Willebrand *et al.*, 2001) to the data-based analysis (Talley *et al.*, 2003) is not unexpected. Both are essentially the result of geostrophic adjustment of climatological data, combined with Ekman transports. None of the models has time to adjust fully to the thermal forcing; remember that the thermohaline circulation takes centuries, if not millenia, to set up, so we don't know what any of these models' estimates of the ocean's climate will be. The differences are perhaps more surprising.

We do not expect models in this configuration to represent intense currents such as the Gulf Stream faithfully, but we do expect them to conform to the Sverdrup balance in the interior (see also Exercise 3.3(b)). Figure 4.30 shows the annual mean cumulative barotropic transport as a function of longitude between the given longitude and the eastern boundary at 27° N. East of 70° W most of the models agree with the value derived from the Sverdrup balance, but MOM shows large deviations in the eastern half of the basin. This result is evidently not robust: in the analogous comparison in Willebrand *et al.* (2001), the model based on MOM is the one that follows the Sverdrup relation most closely, while the ROMS-based model and the layer model deviate from the Sverdrup balance in the eastern part of the basin. These discrepancies are ascribed to interaction with topography and the form of the equation of state respectively.

All of the models will reproduce the essential features of the large-scale circulation of the Atlantic. Given the criteria by which they were compared, among themselves and with available observations, they could

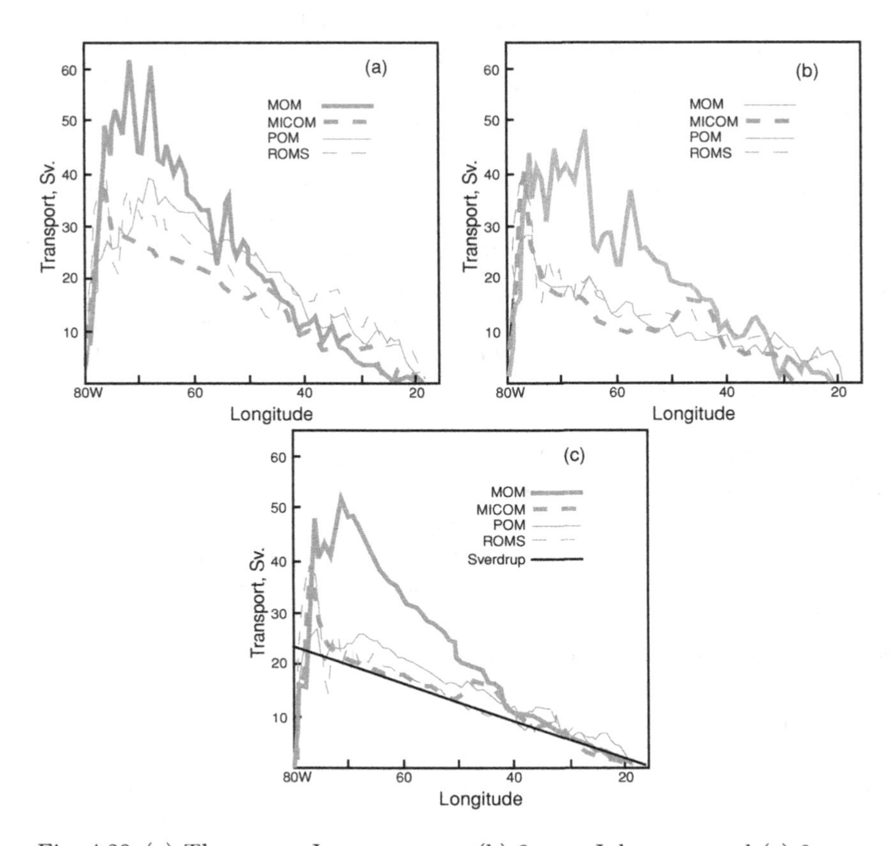

Fig. 4.30 (a) Three-year January mean, (b) 3-year July mean and (c) 3-year annual mean barotropic transport for MICOM, MOM, POM and ROMS at 27° N, along with the transport estimated from the Sverdrup relation. Redrawn from Figure 9 of Chassignet *et al.* (2000), with permission of Elsevier.

hardly do otherwise: most of the results reflect Ekman dynamics near the surface and geostrophic adjustment in the interior. Some results suggest the inherent features of these models: excessive diapycnal diffusion in level models such as MOM, and difficulties in proper representation of pressure gradients in level models such as MICOM that lead in this example to the inability to represent the intrusion of Antarctic Bottom Water northward into the Atlantic basin at depth. There are remedies, at least partial ones, for these problems. For our purposes, we will conclude that results of comparisons such as this one are highly dependent on specific aspects of model configuration.

4.11 Open-boundary conditions

It is often convenient or useful to formulate regional models, in which some or all of the lateral boundaries are not natural boundaries, but rather artificially drawn boundary lines in the ocean. One of the most obvious examples is the case of coastal models, in which the region of interest is the vicinity of the coastline, and the modeler does not wish to spend the resources required for computing model solutions far out into the open sea. In that case, one might wish to consider the seaward boundary as being open, and either impose a Sommerfeld radiation condition (see Section 3.7), or perform a *nested* calculation, in which boundary conditions are derived from a larger-scale model, perhaps one with coarser resolution.

As we have seen in Section 4.4, the primitive equations admit wavelike solutions with speeds that decrease roughly as $O(1/n)$ for the nth vertical mode, so as we increase the vertical resolution of our models, the models will admit wave modes that travel arbitrarily slowly. This is the root of the greatest problem with open-boundary-value problems for the primitive equations. In Section 4.4, we saw that, at an open-boundary, we must specify a number of quantities equal to the number of wave modes that enter the domain. In general, the fastest baroclinic waves are faster than advection speeds commonly found in the ocean, so the open-boundary problem is usually subcritical with respect to those wave modes, but as we have seen, as the vertical grid is refined, the model will eventually admit a wavelike solution slower than the advection speed on the boundary, and the open-boundary problem will be subcritical and supercritical at the same time. There is no obvious way out of this dilemma: for this reason, no locally specified open-boundary condition can lead to a well-posed problem. Details can be found in Oliger and Sundstrom (1978).

Of course, if the velocity normal to the open boundary vanishes identically, then the problem is always subcritical for any wave, but this condition is too restrictive for most applications. There are nonlocal boundary specifications that may yield better behavior, i.e., one can formulate a boundary condition that involves a Fourier transform in the tangential direction. This has been done for certain classes of waves (see, e.g., Bennett, 1976), but implementation is cumbersome. We could also filter in the vertical so the slower waves with the finer scale vertical structure would not appear on the boundary, but this is essentially imposing dissipation, and could lead to artificial boundary layers. In any case, this solution is not commonly tried.

Open-boundary problems for vertical discretizations of the primitive equations can lead to well-posed problems, even though the open-boundary problem for the full three-dimensional primitive equations does not. This is because the discretized system only admits as many vertical modes as it has levels or layers. If the slowest wave speed admitted by the discretized system is still faster than the boundary velocity, then the whole system is subcritical and therefore well-posed. One could imagine examples in which a well-behaved calculation would become unstable upon refinement of the vertical grid.

Most open-boundary value methods used in practice are dissipative. The equations then become parabolic, and a well-posed problem can be formulated by specifying all relevant quantities on the boundary. With sufficient attention to details such as cell Reynolds number limitations (see Section 3.3.1) stable calculations can be performed in this setting, but one must be aware of the possibility of computational artifacts such as boundary layers in the neighborhood of the open boundary.

4.12 Finely resolved calculations

In the last decade, advances in computing have made computing resources available that allow practical calculation of global primitive equation models with eddy-resolving, or at least "eddy-permitting" resolution. By "eddy-permitting" resolution, we usually mean resolution near the local internal deformation radius, but not fine enough to resolve the details of motions on the scale of the internal radius faithfully. Such models exhibit eddies, possibly with useful properties, but they are not faithfully rendered.

The tools of numerical analysis are not by themselves sufficient to rule on the question of "how much resolution is enough?" let alone the larger questions about how to assign quantitative measures of reliability to a finely resolved model. The techniques of numerical analysis are geared more towards deciding whether a given technique, implemented with a given set of parameters, results in an accurate solution of a given equation. Detailed ocean general circulation models contain so many parameterizations and physical approximations that it is difficult to say with certainty precisely which equations are being solved.

All general circulation models require descriptions of the boundary and bottom topography, as well as specification of wind stress and heat flux. Details of geography often call for compromises whose consequences are not easily evaluated. Even those models with grids as fine as $0.25°$ are required to make some compromise at the Strait of Gibraltar, either artificially widening the strait so that it contains at least one interior point, or sealing the strait altogether and making some allowance for salt fluxes into the North Atlantic. Poleward extension of models requires either some explicit model of the influence of sea ice, or some parameterization of its effect. Imposition of artificial walls at the limits of poleward extent of models also requires careful specification. Deficiencies in any or all of these inputs could result in systematic model-data misfits that might even be exacerbated by extremely fine resolution.

Actual convergence, say, in the strict sense of Section 2.2.1 is hard to come by. Schmitz and Thompson (1993) performed a series of runs of a two-layer model of the Gulf Stream in which all parameters but the grid spacing were held constant. Theirs was a regional model, extending from $30 - 45°$ N and $45 - \approx 70°$ W, with transport prescribed at inflow ports south of Cape Hatteras for the Gulf Stream and near the Grand Banks for the deep western boundary current (DWBC). With only two layers in the vertical plane, the model could not be expected to reproduce the details of gulf stream variability faithfully, but their study was one of the few to isolate the effect of resolution. They performed three experiments, with horizontal resolutions of $0.2°$, $0.1°$ and $0.05°$. The only quantities tabulated were energy statistics. Their results are shown in Table 4.2.

Values of energy statistics are shown for different resolutions, along with the value estimated for the exact quantity by Richardson extrapolation. It is interesting to note that values predicted by Richardson extrapolation imply that the difference between the $0.05°$ and the $0.1°$ should have been greater than it was, based on the assumption that the method was in fact second order. Two alternatives are possible: (i) the $0.2°$ run may not be in the asymptotic range for the numerical method, i.e., the coefficient of the lead term in the truncation error expansion may not be constant as the resolution increases; (ii) it is also possible that the calculations of the energies and energy exchanges were not stable beyond a single significant place, i.e., if the simulations had been run longer, these averaged numbers might have changed in the second or third place. In either case it is likely, for this particular setup, that the run with resolution of $0.1°$ is fairly well converged, as the authors claim.

Table 4.2 Results of a resolution study of a regional model of the Gulf Stream. After Table 2 from Schmitz and Thompson (1993)

Quantity	Resolution			
	0.2°	0.1°	0.05°	Extrapolation
TKE1[a]	2.14	3.75	3.67	4.29
TKE2[b]	0.48	0.95	0.92	1.11
MKE1→EKE1[c]	83.3	179.8	184.3	212.0
MPE→EPE[d]	16.7	39.9	43.2	47.6
EPE→EKE2[e]	16.7	39.4	43.2	47.0

a: Total kinetic energy, layer 1
b: Total kinetic energy, layer 2
c: Transfer of mean kinetic energy to eddy kinetic energy, layer 1
d: Transfer of total mean potential energy to total eddy potential energy
e: Transfer of total eddy potential energy to layer 2 eddy kinetic energy

At first glance it seems obvious that comparison to observations should provide the ultimate judgment of the extent to which the output of a model provides a quantitative estimate of the state of the ocean. We now have decade-long time series of remotely sensed observations of the ocean surface, long enough to calculate statistics reliably, without excessive fear that the numbers so derived might be untrustworthy due to the small size of the sample. As valuable as these data sets are for evaluation of large-scale general circulation models, they are also contaminated with errors, and, perhaps more importantly, are the results of extensive processing, which can impart scale-dependent characteristics that cannot be ignored.

Stammer *et al.* (1996) performed a simulation on a Mercator grid with 0.4° resolution in longitude, which works out to an average grid size of 0.25°, and 20 levels in the vertical, with the Parallel Ocean Climate Model (POCM), a descendant of the Bryan–Cox model, as implemented by Semtner and Chervin (1992). Their model was driven by wind-stress fields from the European Center for Medium-Range Weather Forecasting (ECMWF) and imposed heat flux was a combination of monthly heat flux fields from ECMWF (Barnier *et al.*, 1995) and restoration to the Levitus sea surface temperature (SST) climatology (Levitus, 1982) by a Newton cooling law with a 30-day timescale. Spinup was accomplished by an elaborate process, beginning with the spinup of a 0.5° version of the model for 33 years starting from the Levitus (1982) temperature and salinity distributions. The final version of the model, designated

"POCM_4B," was run over the period 1987 through 1994. Extensive comparisons were performed with two years of altimeter data, and with World Ocean Circulation Experiment (WOCE) hydrography. Estimates of mean sea level height, as opposed to temporal differences, from altimeter data, were accomplished by using an estimated geoid. Regions north of 58° N and south of 68° S were restored to the Levitus (1982) climatology with a timescale of 120 days. A similar restoration to climatology was imposed near Gibraltar, since the Mediterranean Sea was not explicitly included.

As we have seen in Section 4.4, we would not expect much evolution of the thermohaline circulation, which remains close to the Levitus climatology. In the upper ocean POCM_4B reproduces most of the major current system in reasonable detail, but the temperatures are mostly too cool, by as much as 2°, and the model seems to underestimate variability systematically; the eddy kinetic energy is too small in most parts of the ocean, by a factor of 2–4. This lack of variability can, to some degree, be ascribed to lack of an adequate mixing model for the upper ocean. Other studies, as we shall see presently, indicate that 0.25° resolution is still too coarse to represent variability accurately.

McClean *et al.* (1997) performed a series of comparisons of POCM_4B with a similar implementation with average resolution of 1/6°. They found that even this fairly small increase in resolution resulted in increased variability. In comparisons of their model output with altimetric data, it is important to note the limitations of the observations. McClean *et al.* note that the processing of the data tends to remove energy at scales shorter than about 60 km, and the processing necessary to remove tides from the altimeter signal tends to cast suspicion on model–data comparisons for oscillation periods corresponding to 40–60 days and 20–30 days.

One might reasonably ask whether the improvement from increased resolution comes from more accurate description of smaller-scale motions, or from the decreased dissipation allowed by the finer-resolution models. Direct comparison would require running the finer-resolution model at the dissipation level of POCM_4B. Rather than going to the expense of performing such an experiment, McClean *et al.* note that both models resolve the local scales of motion at low latitudes. If dissipation were the deciding factor, the greater dissipation of POCM_4B would lead to less variability, but the variability of the two models is about the same at low latitudes. This result weighs in on the side of finer resolution being responsible for the increased variability. McClean *et al.* conclude that even the 1/6° resolution is probably not enough.

Maltrud *et al.* (1998) applied three different forcing methods to the 1/6° model used by McClean *et al.* All were driven by ECMWF winds and surface salinity was relaxed to climatology (Levitus, 1982) with a timescale of one month. In the simulation with the least detail, which they call "POP5," monthly average winds were imposed, surface temperature and salinity were relaxed to climatology (Levitus, 1982) with a timescale of one month, and the Mediterranean was closed off from the Atlantic at Gibraltar. "POP7," the run performed at the next level of detail was similar, but was forced with three-day average winds, rather than monthly average winds. In "POP11," the most complex of the three simulations, the heat flux was derived from surface analysis of the ECMWF fields as described by Barnier *et al.* (1995) and the Mediterranean was included by widening the strait of Gibraltar to admit an interior gridpoint. These differences in forcing made a considerable difference in the final results. Mean current sections at 26.5° N eastward from 77° W, off Abaco, in the Bahamas are shown in Figure 4.31(a) and (b). The Gulf Stream has a slightly greater maximum and a tighter core in POP11, and the difference in the deep western boundary currents is even more striking, being much stronger and extending much deeper in POP11. Results from POP7 and POP11 can be compared to similar results obtained by Smith *et al.* (2000) in their 1/10° simulation of the North Atlantic. Their velocity section off Abaco is shown here in Figure 4.31(c). The result from the finer-resolution story shows the Gulf Stream and the deep western boundary current to be stronger still, with the core of the deep western boundary current deeper, around 2400 m, as opposed to about 1700 m for POP11. A similar section, drawn from mooring data, is shown in Figure 4.31(d). The observed circulation clearly extends deeper than in the POP7 and POP11 runs, in closer agreement with the 1/10° run, the current maxima are closer to the boundary and the transport is closer to the estimated value of 40.5 Sv. The influences that lead to a deeper deep western boundary current can not be clearly isolated from existing runs at this time. The issue is further confused by the result of an even coarser resolution run, with 0.4° zonal resolution, 1/3° meridional resolution and 30 levels in the vertical plane, which exhibits the deep western boundary current near the depth expected from the current meter data and the 1/10° run, although the current in the model is too slow (Böning *et al.*, 1991).

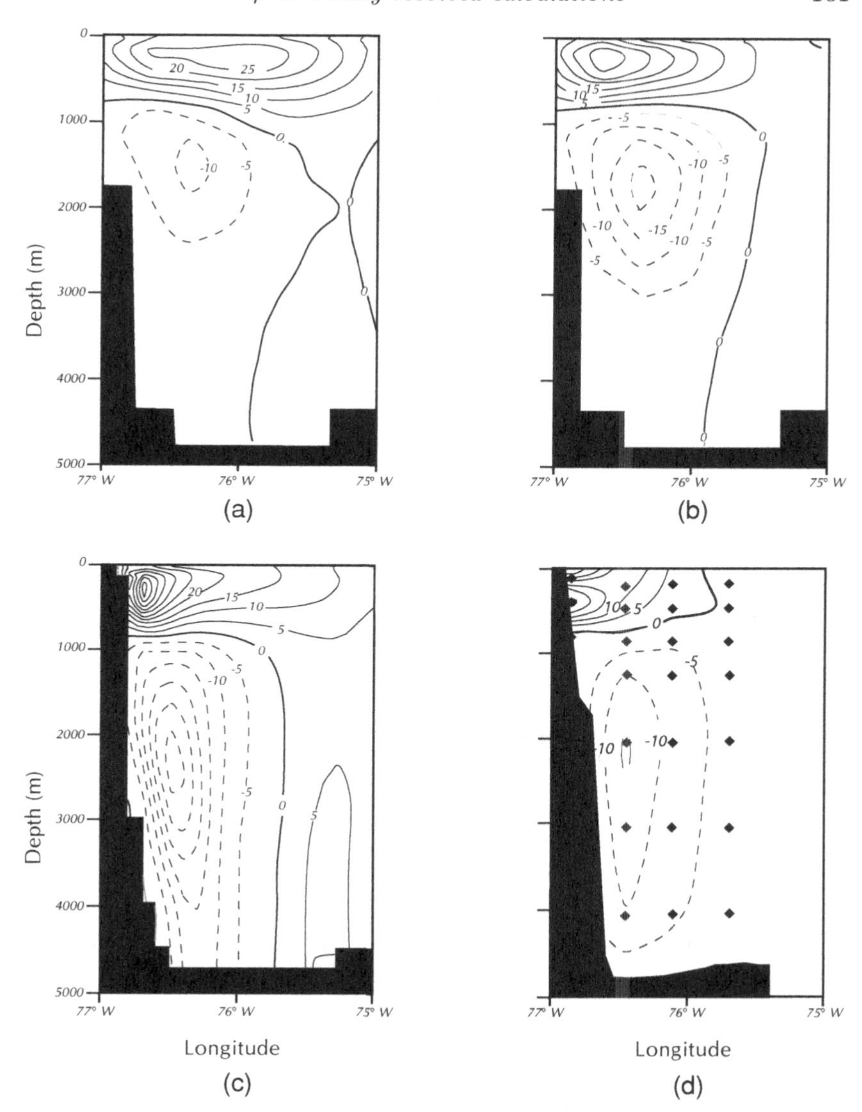

Fig. 4.31 Cross-section of meridional velocities at 26.5° N: (a) POP7, (b) POP11, (c) 0.1° run. (d) Mean meridional velocity section from a moored array, June 1990–February 1992. Instrument locations shown as heavy black diamonds. Redrawn from Figure 5 from Maltrud *et al.* (1998), Figure 14 of Smith *et al.* (2000) and Figure 3 of Lee *et al.* (1996), with permission of the American Meteorological Society and the American Geophysical Union.

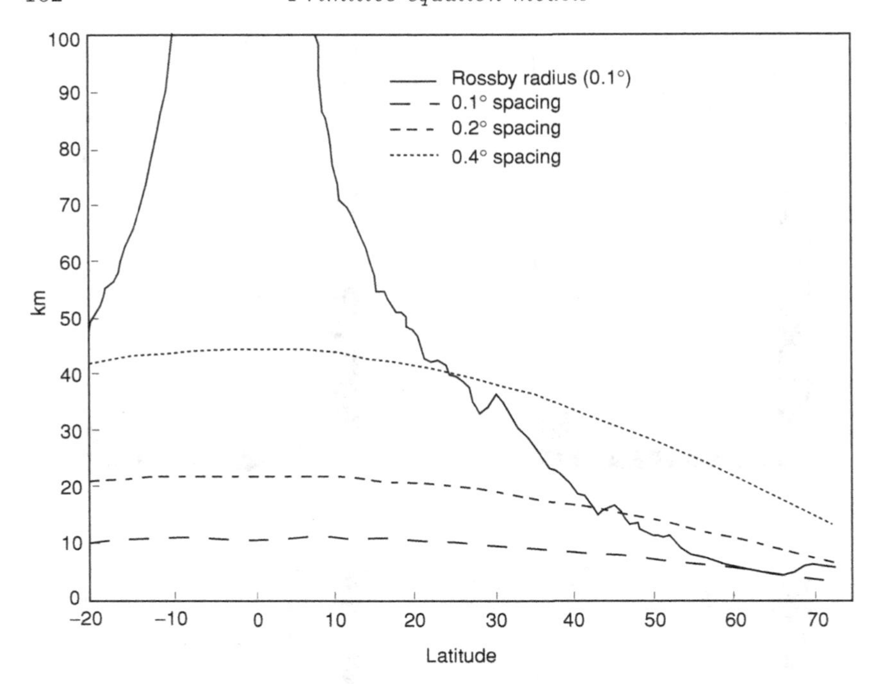

Fig. 4.32 Zonally averaged radii of deformation, plotted on the same axis with grid spacing for 0.4°, 0.2° and 0.1° runs (Bryan *et al.* (1998)). Based on Figure 1 of Smith *et al.* (2000), with permission of the American Meteorological Society.

The model currents may actually be too swift. We cannot know for sure, since the resolution of the mooring array is rather coarser than 1/10°, and we have the added complication of variability that can affect the averaging process.

While there are still significant systematic errors in the 1/10° simulations, it seems that a line has been crossed in the transition from 1/6° to 1/10°. In the fine-resolution simulations, the chronic problem of northward overshoot of the Gulf Stream separation seems to have been fixed, and the variability statistics are no longer systematically low. The essential point may be resolution of the internal deformation radii. Figure 4.32 shows the zonally averaged deformation radius as a function of latitude, calculated from a 0.1° simulation, plotted on the same axes as the gridspacing, in km for grids of 0.4°, 0.2° and 0.1°. At this point the literature does not contain sufficient information to decide whether the major source of error in the most finely resolved simulations is still lack of resolution, or some feature of the forcing, the boundaries or some parameterizations.

The new finer-resolution studies are extremely demanding in terms of computing resources. This applies to both the running of the model and the processing and analysis of the output, which can easily run to the better part of a terabyte.

Coarser-resolution models, i.e., those with a horizontal resolution of a degree or more, will continue to have a place in ocean modeling, despite the evident fidelity of the high-resolution models to the observations, if only for reasons of resource conservation. Coupling of atmospheric or biogeochemical models to the finest resolution models is still prohibitively expensive as this is being written. One day soon, this may no longer be the case. When that day comes, we will see an even greater emphasis on efficient storage, processing, analysis and visualization methods.

4.13 Exercises

4.1 Consider the shallow-water system in the half-plane $-\infty < x \leq 0$, with the boundary condition $u = 0$ at $x = 0$. This problem admits the *Kelvin waves*, a family of waves that propagate along the boundary with an amplitude which decays exponentially away from the boundary. It might seem from (4.23) that such a family of waves might exist in the shallow-water system with a rigid lid. Show that in fact they do not, i.e., the rigid lid suppresses the Kelvin waves.

4.2 Derive the approximation in (4.26) from (4.24). Apply the approximation $U - c \approx 0$ to derive c_R and $U - c \approx (\Psi_0 + f^2/k^2)^{1/2}$ to derive $c_\pm$. Suggest a physical interpretation for your results.

4.3 Consider a density profile in which the density is independent of the horizontal directions and depends linearly on z. Assume that the horizontal resolution is sufficiently fine that the topography is well represented by a linear function, i.e., $D(x) = H_0 - sx$. Show that, in this case, (4.92), (4.93) and (4.94) are exactly correct, i.e., the pressure gradient vanishes identically.

4.4 If the background density field in a given model simulation is perturbed by a disturbance that depends on z alone, the two terms in (4.89) must be comparable in magnitude. Show that, in this case, both terms will be proportional to $g\delta A/L$, where A is the characteristic amplitude of the topography and L is the horizontal scale of the topography. Therefore deduce that if a second-order method is used to approximate (4.89), the truncation error will be proportional to $g\delta(A/L)(\Delta x/L)^2$. This is the estimate given by McCalpin (1994).

4.5 Write two codes to solve the nonlinear shallow-water equations in one dimension, one with the leapfrog scheme and the other with the Lax–Friedrichs scheme. Use your codes to solve the dam-break problem, i.e., apply the following initial conditions: $u = 0$ on the entire domain, and $h = H_0$ for $x \leq 0$ and $h = 0$ for $0 < x$. Compare the results from the two schemes to one another and to the exact solution (see, e.g., Whitham, 1974).

5
Quasigeostrophic models

5.1 Background and notation

The quasigeostrophic model is a simplification of the primitive equation model that retains many of the essential features of geophysical fluid flow. Details of the mathematical and physical approximation may be found in Pedlosky (1979). The computational treatment here follows Miller *et al.* (1983).

The Rossby number R_0 in the case of interest is very small, i.e., the Coriolis timescale is much faster than the advective timescale or, equivalently, the scale of the inertial acceleration U^2/L is small compared to that of the Coriolis acceleration Uf_0. In order to derive the most basic quasigeostrophic model, we must assume that $\beta L/f_0 \sim R_0$. This assumption is often violated in practice, but we need it here in order to guarantee that the divergence of the geostrophic velocity field vanishes. We write the geostrophic relations:

$$u = -\psi_y, \quad v = \psi_x, \quad P = \rho_0 f_0 \psi. \tag{5.1}$$

If we scale u and v by U, and x and y by L, we find $P = \rho_0 f_0 UL\psi$. We scale time according to the Rossby wave timescale $T = 1/\beta L$. Hereafter, we shall assume that all quantities are in nondimensional form, unless otherwise noted.

The geostrophic relations give us no guidance as to the evolution of the velocity field; a detailed derivation of the now-classical result that the field evolves in such a way as to conserve potential vorticity can be found in Pedlosky (1979). The relevant equations are

$$\frac{\mathrm{D}}{\mathrm{D}t}\left(\zeta + \frac{f}{\epsilon}\right) = F \equiv \text{forcing} + \text{dissipation}, \tag{5.2}$$

$$\nabla^2\psi + \Gamma^2(\sigma\psi_z)_z = \zeta, \tag{5.3}$$

$$\frac{\mathrm{D}}{\mathrm{D}t} = \frac{\partial}{\partial t} + \epsilon J(\psi, \cdot), \tag{5.4}$$

$$J(\psi, \cdot) = \psi_x \frac{\partial}{\partial y} - \psi_y \frac{\partial}{\partial x}, \tag{5.5}$$

$$\epsilon = \frac{U}{\beta L^2}, \tag{5.6}$$

$$f = f_0 + y, \tag{5.7}$$

$$\Gamma^2 = \frac{f_0^2 L^2}{N_0^2 h_T^2}, \tag{5.8}$$

$$\sigma = \frac{N_0^2}{N^2(z)}, \tag{5.9}$$

$$N_0 = \text{buoyancy frequency scale}, \tag{5.10}$$

$$N = \text{buoyancy frequency} = \left(-g\frac{\partial\bar{\rho}}{\partial z}\right)^{1/2}, \tag{5.11}$$

$$h_T = \text{depth scale}, \tag{5.12}$$

$$\rho = \text{density} = \rho_0\left[1 + \bar{\rho}(z) + \left(\frac{f_0 UL}{g h_T}\right)\delta(x, y, z, t)\right]. \tag{5.13}$$

(Note that we have scaled (5.6) so that $\beta = 1$.) The quantity $\zeta + f/\epsilon$ is the potential vorticity. The parameter ϵ may be viewed as the ratio of the Rossby wave timescale $1/\beta L$ to the advective timescale L/U. This parameter therefore measures the relative weight of the nonlinear term. ϵ is sometimes known as the β-Rossby number. δ is the dynamic density anomaly. With this choice of scale, the hydrostatic relation for the dynamic pressure becomes $\psi_z = -\delta$. The flow is assumed to satisfy the Boussinesq and adiabatic approximations.

This model does not contain gravity waves since the horizontal divergence of the velocity vanishes. This allows much longer time steps than would be permitted in primitive equation models when explicit methods in time are used.

The quasigeostrophic framework restricts the slope of the isopycnals and the slope of the bottom to $O(R_0)$. This does not stop people from violating this condition, but usually they feel guilty about it. Suppose the bottom relief is given by $z - H(x,y) = G(x,y,z) = 0$. The boundary condition at the bottom is given by

$$(u, v, w) \cdot \nabla G = w - uH_x - vH_y = w - J(\psi, H) = 0 \tag{5.14}$$

(recall that $J(\psi, H) = \psi_x H_y - \psi_y H_x$).

A higher value R_0 implies that greater slopes are allowed but the approximation is less accurate since it is based on an expansion in powers of R_0.

5.2 Computation

The equations of motion may be written

$$\zeta_t + \epsilon J(\psi, \zeta) + \psi_x = F, \tag{5.15}$$

$$\nabla^2 \psi + \Gamma^2 (\sigma \psi_z)_z = \zeta. \tag{5.16}$$

The key point is that z appears only as a parameter in (5.15) and (5.16) is linear. It is therefore easy to construct level models since the levels are coupled only through the vortex stretching terms in (5.16), so the nonlinear equation has only two space dimensions. The following simple procedure can therefore be used: given initial values ψ_0, ζ_0 for ψ and ζ at $t = 0$ at each level, calculate ζ at time Δt using (5.15). This is a purely 2-D calculation at each level. Now that ζ at time Δt is determined, use this new value of ζ to calculate new values for ψ from (5.16). Because fully 3-D elliptic solvers are rare, it is common to use separation of variables to solve (5.16).

For the advection operator, there are methods that conserve ζ^2 and $(\nabla\psi)^2$. One is the Arakawa scheme. Details of the construction of the Arakawa Jacobian are given in Haltiner and Williams (1980, Chapter 5); see also Haidvogel *et al.* (1980). The bilinear finite-element model gives the identical form of the advection operator to the Arakawa scheme (Jespersen, 1974).

5.2.1 Vertical discretization

For the vertical operator, let

$$\psi_j(x, y, t) = \psi(x, y, z_j, t),$$

where z_j is the (dimensionless) vertical coordinate of the jth level.

We write the finite-difference operator as follows:

$$(\sigma\psi_{1z})_z \cong \frac{1}{h_1}\left\{\sigma\psi_z\big|_{\text{top}} - \frac{\sigma}{h_2'}(\psi_1 - \psi_2)\right\}, \tag{5.17}$$

$$(\sigma\psi_{jz})_z \cong \frac{1}{h_j}\left\{\sigma_{j-1}\frac{\psi_{j-1}-\psi_j}{h_j'} - \sigma_j\frac{\psi_j-\psi_{j+1}}{h_{j+1}'}\right\}, \quad 0 < j < N, \tag{5.18}$$

$$(\sigma\psi_{Nz})_z \cong \frac{1}{h_N}\left\{\sigma_{N-1}\frac{\psi_{N-1}-\psi_N}{h_N'} - \sigma\psi_z\big|_{\text{bottom}}\right\}. \tag{5.19}$$

The h_j and h_j' represent the depths at which the computed quantities are calculated; a diagram of the vertical discretization scheme appears in Figure 5.1. $\sigma\psi_z\big|_{\text{top}}$ and $\sigma\psi_z\big|_{\text{bottom}}$ represent density anomaly distributions at the top and bottom boundaries which are determined prognostically through the relation between vertical velocity and density advection. This follows from the equation for adiabatic flow. As noted above, the vertical velocity at the bottom can be determined by the kinematic boundary condition. The vertical velocity at the top could come from an Ekman layer type calculation. The simplest cases studied here correspond to uniform temperatures at the top and bottom. If there is no surface forcing or bottom relief, we have $\sigma\psi_z\big|_{\text{top}} = \sigma\psi_z\big|_{\text{bottom}} = 0$.

If we let $\psi = (\psi_1, \psi_2, \ldots, \psi_N)^{\mathrm{T}}$, we may write our discrete approximation to (5.17)–(5.19) in vector form:

$$(\sigma\psi_z)_z \cong L\psi + \begin{pmatrix} \sigma\psi_z\big|_{\text{top}} \\ 0 \\ 0 \\ \vdots \\ 0 \\ \sigma\psi_z\big|_{\text{bottom}} \end{pmatrix}. \tag{5.20}$$

The eigenvalues $-\tilde{\lambda}_i^2$ of L approximate the eigenvalues $-\lambda_j^2$, $j = 0, 1, \ldots N-1$ of the continuous operator $(\sigma\psi_z)_z$ with boundary conditions given by $\sigma\psi_z\big|_{\text{bottom}} = \sigma\psi_z\big|_{\text{top}} = 0$ for an N-level model. This is essentially the same eigenvalue problem as in (4.64)–(4.66). Physically, λ_j is the reciprocal of the jth internal deformation radius. Since the rows of L all sum to zero, $\tilde{\lambda}_0^2 = \lambda_0^2 = 0$ and thus the discrete system, as in the continuous system, has a barotropic mode. However, for the baroclinic modes, the approximation $-\tilde{\lambda}_j^2$ to $-\lambda_j^2$ deteriorates with increasing j.

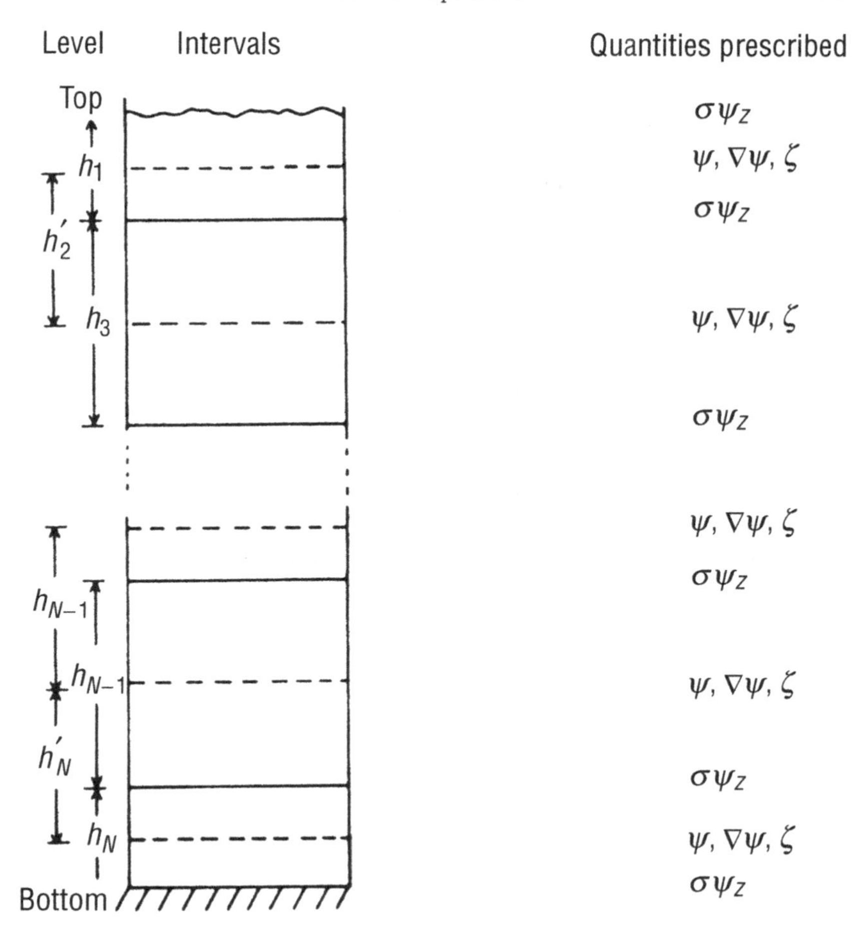

Fig. 5.1 Structure of depth discretization of baroclinic model. Reproduced from Figure 1 of Miller *et al.* (1983), with permission of Elsevier.

For fixed j, $-\tilde{\lambda}_j^2$ converges to $-\lambda_j^2$ as the number of levels N increases; but for any finite N, $-\tilde{\lambda}_j^2$ will be a very poor approximation of $-\lambda_j^2$ for $j \approx N$. This can be illustrated with the simple example of a model depth structure with $\sigma = \Gamma^2 = 1$, total depth $= \pi$, and equally spaced levels. In this case, the eigenvalues of both the continuous operator and the finite-difference approximation are known exactly. The comparison between the exact and approximate eigenvalues is shown in Table 5.1. As the number of levels increases, $-\lambda_1^2$ approaches 1, as shown in row 2 of Table 5.1. But looking down the column corresponding to the

Table 5.1 Eigenvalues of vertical finite-difference operator. Levels are equally spaced. Basin is assumed to have total depth π. After Table III of Miller *et al.* (1983), with permission from Elsevier.

	Number of levels			
Mode	2	4	8	Continuous operator
0 (barotropic)	0	0	0	0
1	−2	−0.95	−0.99	−1.0
2		−3.24	−3.80	−4.0
3		−5.53	−9.01	−9.0
4			−12.97	−16.0

eight-level simulation, we see the error increase from 1% for the first mode to 19% for the fourth. A 16-level example would show a marginally better value for $-\tilde{\lambda}_1^2$, and a considerably better value for $-\tilde{\lambda}_4^2$, but the error in $-\tilde{\lambda}_{15}^2$ would be high. In order to examine the effect of truncation error in the vertical directs, consider the linear equation

$$\zeta_t + \psi_x = 0, \tag{5.21}$$

$$\nabla^2\psi + (\sigma\psi_z)_z = \zeta. \tag{5.22}$$

Then, separate variables to get

$$\psi = \sum_j \hat{\psi}(x, y, t) Z_j(z), \tag{5.23}$$

$$(\sigma Z_j')' = -\lambda^2 Z_j. \tag{5.24}$$

Next, look for solutions of the form

$$\psi_j = \Psi \mathrm{e}^{\mathrm{i}(kx+ly-\omega t)}, \tag{5.25}$$

$$(-k^2 - l^2 - \lambda_j^2)\Psi = \zeta, \tag{5.26}$$

$$-\mathrm{i}\omega(-k^2 - l^2 - \lambda_j^2)\Psi + \mathrm{i}k\Psi = 0. \tag{5.27}$$

Thus

$$\omega = \frac{-k}{k^2 + l^2 + \lambda_j^2}, \tag{5.28}$$

$$\lambda_j^2 = \frac{1}{(r_j^2)}, r_j \equiv \text{the } j\text{th internal radius}. \tag{5.29}$$

If a finite-difference operator is used to estimate the vertical derivatives, the truncation error in the vertical direction will appear as error in the wave speeds.

Table 5.2 Response of Shapiro filters of orders 1, 2 and 4. Values shown are factors by which the wave with the given wavelength is attenuated. The wavelength is given in gridpoints. After Table II of Miller *et al.* (1983), with permission from Elsevier.

Order			
1	2	4	Wavelength
0.0000	0.0000	0.0000	2
0.5000	0.7500	0.9375	4
0.7500	0.9375	0.9961	6
0.9330	0.9955	1.0000	12

5.3 Dissipation

In practical computation, some dissipation must be imposed. Often bottom friction is used in the form $F = -\kappa\zeta$ for some κ, but that expression is not scale selective. Imposition of an eddy viscosity often damps large-scale features more than we would like. Therefore, high-order friction ($\sim \nabla^6, \nabla^8, \nabla^{16}$) and so forth is often used. The advantage of high-order friction is that it damps the shortest waves, i.e., those with wavelength $2\Delta x$, while not having so much effect on those with longer wavelengths. Table 5.2 shows the effect of high-order friction, implemented here in the form of the Shapiro (1970) filter. In this implementation, the parameters are chosen so that the wave with wavelength $2\Delta x$ is exactly annihilated in a single timestep. Note that the first-order filter, which is equivalent to ordinary Laplacian friction, damps the wave with wavelength $6\Delta x$ by 0.75 each timestep. This wave is actually fairly well resolved, and we would not like it so strongly damped. The second order filter, equivalent to ∇^4 friction, damps this wave by 0.94, a considerable improvement. This is sometimes known as "biharmonic friction," and is a common choice. The fourth-order filter leaves the wave nearly intact.

5.4 Open-boundary models

It is often desirable to model detailed regional dyanamics with models that do not refer to physical boundaries. This can have the advantage of concentrating computing resources in a region of interest, which may be far from the nearest coastline. This can be the most efficient way to achieve detailed resolution of eddy phenomena; see, e.g., Robinson and Walstad (1987).

For primitive equation models, we saw in Section 4.11 that no local boundary condition specifies a well-posed problem (but see Spall (1988) for a slightly different point of view). For quasigeostrophic models, the technical picture is less clear. Recall that the equations specify that a scalar quantity, the potential vorticity ζ, is conserved along particle paths. Therefore, Charney *et al.* (1950) proposed that ζ be specified at inflow points on the boundary and be determined from upstream at outflow. This can lead to problems in a simply connected domain, where there will be points at which the flow is tangent to the boundary, and inflow and outflow points will be adjacent to one another. This can lead to arbitrarily large potential vorticity gradients. Elementary examples can be found in Miller and Bennett (1988).

Several approaches have been presented for the actual implementation of open boundary problems for the quasigeostrophic model. The problem is that the computation requires some specification of the potential vorticity at outflow points. Haidvogel *et al.* (1980) compared several computational schemes for solving this problem. The simplest thing to do is to extrapolate ζ from the interior. This results in an unstable scheme. The next step would be to form a space–time interpolation scheme. Sundstrom (1969) and Davies (1973) proposed the following:

$$\zeta_B^k + \zeta_{B-2}^k = \zeta_{B-1}^{k+1} + \zeta_{B-1}^{k-1}, \tag{5.30}$$

where the subscript "B" denotes a value at a boundary point, and the subscript "$B-k$" denotes an interior point k gridpoints from the boundary in the direction normal to the boundary. Superscripts denote the timestep. This can be viewed in one of three ways: (i) The scheme equates time and space averages of ζ at the point $B-1$. From this point of view, the scheme imposes a smoothness condition. (ii) Upon subtracting $2\zeta_{B-1}^k$ from both sides, the scheme appears as an approximation to $\zeta_{tt} = c^2\zeta_{xx}$, a sort of local wave equation with $c^2 = \Delta x^2/\Delta t^2$. (iii) The scheme can be viewed as a low-order extrapolation scheme.

Haidvogel *et al.* (1980) also considered another space–time extrapolation scheme:

$$\zeta_B^k = 2\zeta_B^{k-1} - \zeta_{B-2}^{k-2}.$$

This scheme, which is due to Kreiss, is discussed in Orlansky (1976).

Haidvogel *et al.* (1980) proposed an open-boundary scheme for use with a finite-element formulation. That scheme makes use of the structure of the finite-element equations to separate the interior equations from those on the boundary. They derived a prognostic equation for the

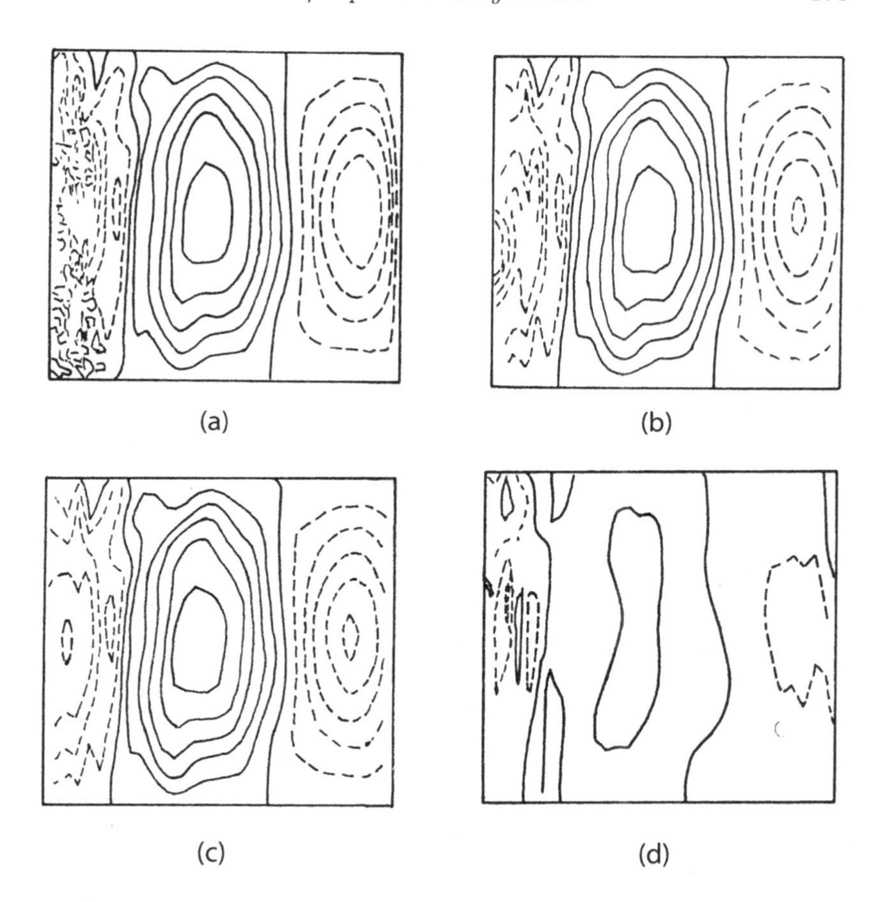

Fig. 5.2 Results of quasigeostrophic open-boundary calculations. Comparison of computational methods for implementation of open-boundary conditions. Each panel shows the deviation of the vorticity field from a reference solution for the nonlinear box mode for a different computation scheme. (a) Boundary vorticity overspecified, (b) space–time extrapolation (Kreiss), (c) Sundstrom–Davies, (d) finite element. Redrawn from Figures 5 and 6 of Haidvogel *et al.* (1980), with permission of Elsevier.

boundary vorticity in terms of the interior vorticity, allowing the total prognostic calculation to be done in two steps.

In Figure 5.2, the three finite-difference methods exhibit roughly the same spatially averaged RMS error, but the Sundstrom–Davies condition produces less small-scale noise at the boundary. The finite-element formulation is better than any of the three others.

Examples of a comparison between results of primitive equation models and quasigeostrophic models can be found in Semtner and Holland (1978), Spall (1988) and Spall and Robinson (1990). Spall and Robinson (1990) contains a detailed description of an experiment in which the quasigeostrophic model of Miller *et al.* (1983) and a primitive equation model based on that of Bryan and Cox (1967) were used to forecast the evolution of the Gulf Stream meander and ring system between Cape Hatteras and the Grand Banks. The domain was a rectangle oriented at a slight angle relative to the parallels of latitude, contained in a region extending from roughly 45° W to 65° W and 33° N to 45° N. This is a particularly stringent test for a quasigeostrophic model. Not only does the swift Gulf Stream flow imply a fairly high Rossby number, but the region includes the New England Seamount Chain, with slopes of the order of 0.1. This simulation is therefore at the very outer limit of the parameter range in which the quasigeostrophic model is expected to do well.

Even though (with a few exceptions) ocean dynamics are well-described by the quasigeostrophic model, application of quasigeostrophic models has been largely confined to process studies in recent years. An example of the use of a quasigeostrophic model for a study of nonlinear basin-mode interaction in the spirit of the examples reported in Section 3.8.2 can be found in Ben Jelloul and Huck (2003). Hogg *et al.* (2003) describe a coupled model of the ocean and atmosphere in which both the oceanic and atmospheric components are modeled by quasi-geostrophic dynamics. Extensive justification was given for this choice.

Several reasons can be advanced for the recent lack of popularity of these models. Surely the ready availability of full-featured primitive equation codes on the internet has something to do with it, but that is not the only reason. Most of the demand on computing resources posed by quasigeostrophic models comes in the solution of the elliptic equation (5.16). Elliptic equations are not so naturally adaptable to parallel computing architectures as are the shallow-water or primitive equations, so the natural computational advantages of quasigeostrophic models are compromised by the fact that, in practice, they are often no more efficient than primitive equation models. Parallel algorithms are being developed for elliptic equations, and when convenient parallel elliptic solvers become widely available and when the ocean modeling community gains experience with them, the natural advantages of quasi-geostrophic models may well lead to increased application.

6
Models of the coastal ocean

6.1 Introduction

In recent years interest in the coastal ocean has increased throughout the world scientific community. Knowledge of the coastal oceans is directly relevant to issues of resource management and security, among others, and is therefore of broad societal interest, since a large proportion of the world's population lives near coastlines. In a purely scientific context, new instruments such as surface velocity mapping radars have been developed, and advances in computers and computing techniques have made detailed models of the coastal ocean practical.

The essential physical mechanisms that determine the most interesting aspects of coastal flow differ from season to season and from location to location. Coastal flows are affected by tides. Nonlinear effects can include rectification, so periodic tidal forcing of the coastal ocean can lead to significant residual steady flows. Interaction of the barotropic tide with topography can result in significant baroclinic motion. Interaction with motions on longer timescales can be nontrivial, and simply averaging over tidal periods or implementing other strategies for filtering out the relatively high frequency tidal motions may not be sufficient to deal with tidal interactions. River outflow, with its associated buoyancy fluxes, can be important.

Coastal upwelling, with its ecological implications, is an important feature of coastal circulation in many areas, as are the coastal jet and the ubiquitous coastally trapped waves, which propagate with the coast on their right as you face in the direction of propagation in the northern hemisphere (see Exercises 6.2 and 6.3).

Surface and bottom boundary layers are of the order of tens of meters thick, and often contain many of the phenomena of interest, beyond simple vertical transfer of momentum and heat. Their structure cannot be ignored, and coastal models must deal explicitly with dissipation. For this reason, most coastal models include a turbulence model.

Faithful models of the coastal ocean must account for detailed topography. For this reason the use of terrain-following coordinates such as σ- or s-coordinates is nearly universal.

Given favorable winds, upwelling and the coastal jet are robust features of many physical models of the coastal ocean. The first theory was proposed by Charney (1955). Charney's theory was based on a quasigeostrophic model, a schematic one to be sure, but one capable of producing upwelling and the coastal jet. Charney's quasigeostrophic model contains a basic mechanism for some of the essential features of coastal flow but does not give a faithful picture of its details.

Theories of coastal flow, including upwelling, the coastal jet and coastal waves are well developed and have been covered extensively in the literature. Prospective modelers of the coastal ocean are strongly encouraged to read the excellent reviews by Brink (1991) and Allen (1980).

Simple linear wave theories such as those found in Brink (1991) have been applied successfully to the wind-driven variability of coastal sea level and current. Variability on timescales of a few days and longer and spatial scales of a few hundred kilometers are reasonably well simulated by a simple model in which the pressure anomaly P is assumed to take the form

$$P(x, y, z, t) = F(x, z)\phi(y, t), \tag{6.1}$$

where x, y and z are the cross-shore, alongshore and upward directions respectively. F is the solution to an eigenvalue problem; see Exercise 6.3 and Brink (1991). ϕ evolves according to an inhomogeneous advection equation,

$$\phi_y + c^{-1}\phi_t + a\phi = b\tau, \tag{6.2}$$

where c is the wave speed from the eigenvalue problem, a is the friction damping coefficient, which must be small in order for this model to work, and can be determined empirically, τ is the wind stress and b is the coefficient that couples the wind stress to the amplitude of the wave response. Equation (6.2) can be solved by integrating along characteristics, similar to the way in which (2.1) was solved. This theory is quite successful within its limits; see, e.g., Battisti and Hickey (1984) and Halliwell and Allen (1984) for application of this theory to the west coast of North America. Other examples can be found in the references in Brink (1991).

Detailed simulation of the coastal ocean, including, e.g., upwelling, downwelling, the behavior of intense jets and residual effects of tides, requires more detailed, complex and computationally demanding models. We next examine a few modeling studies in detail.

Since we expect the alongshore scales of subinertial coastal motion to be much greater than the cross-shore scales, it seems natural to construct two-dimensional models in which derivatives in the alongshore direction are neglected. We expect such models to be able to capture many of the details of the upwelling circulation, and the two-dimensional formulation should allow us to perform efficient calculations with very high resolution. Examples of such calculations can be found in Allen *et al.* (1995) and Federiuk and Allen (1995). Some of these calculations are described in detail in the next section.

6.2 Example: A high-resolution two-dimensional model of upwelling circulation on the Oregon shelf

In these purely two-dimensional calculations derivatives are assumed to vanish in the alongshore direction, hence this model cannot support coastal waves. The model is an implementation of the Princeton Ocean Model (POM; cf. Blumberg and Mellor, 1987), which is formulated, as are nearly all coastal models, in σ-coordinates. The equations of motion are implemented on a C-grid. The idealized coastal geometry is shown in Figure 6.1. The horizontal extent of the model is limited by an artificial wall at $x = 0$, 100 km offshore. The minimum depth at the eastern boundary of the region is 10 m, and the total depth is artificially limited to 500 m. This might seem shallow at first glance, but it is not so shallow as to limit the physical applicability of the calculation, as we shall see, and it allows the authors to avoid even more steeply sloping σ-surfaces with their attendant problems. Several different dissipation schemes are tested in these experiments, the fundamental one being the Mellor and Yamada (1982) level-2.5 turbulence closure.

The upwelling-favorable (in this case southward) wind increases in strength sinusoidally in two days, i.e., $\tau^{(y)} = \tau^0 \sin(t\pi/2T_R)$, $t \leq T_R$; $\tau^{(y)} = \tau^0$, $t > T_R$, $T_R = 2\,\text{days}$. Several different experiments were performed, with different parameter sets. In the basic case (BC), $\tau^0 = -0.05\,\text{Nt}\,\text{m}^{-2}$. Another experiment was performed with twice that magnitude of wind stress.

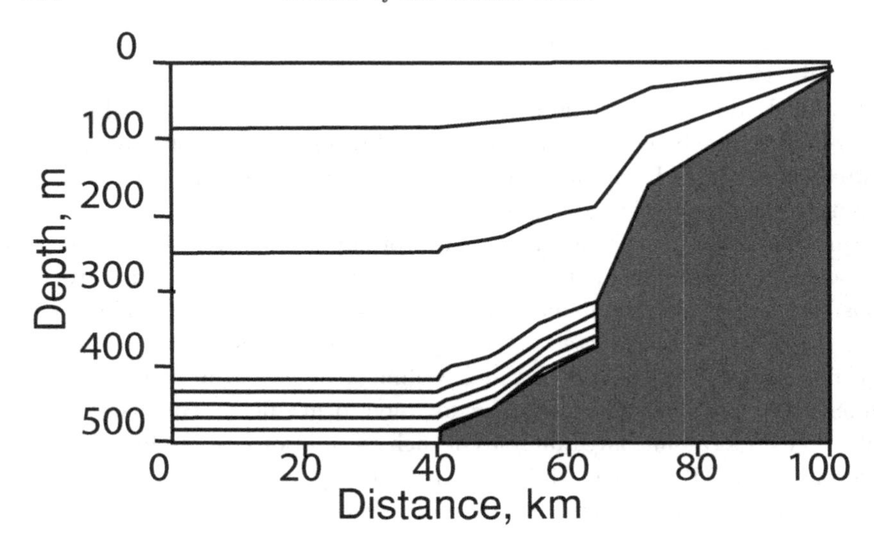

Fig. 6.1 Model computational domain from Allen *et al.* (1995). Solid lines are representative lines of constant σ. Redrawn from Figure 1 of Allen *et al.* (1995), with permission of the American Meteorological Society.

The cross-shore grid spacing $\Delta x = 500\,\mathrm{m}$ and there are 60 equally spaced σ-levels. The condition for hydrostatic consistency (cf. (4.98), Section 4.9) is

$$\left|\frac{\sigma_j D_x}{D}\right| \Delta x \le \Delta\sigma. \tag{6.3}$$

In this case, when $\sigma \approx 0.5$, $D_x \approx 200\,\mathrm{m}/10\,\mathrm{km}$ and D is between 180 and 380 m, hydrostatic consistency is close to the limit. Other experiments were performed with finer horizontal resolution, $\Delta x = 250\,\mathrm{m}$ and $\Delta x = 170\,\mathrm{m}$, and with steeper and less steep bottom slopes, with little change in the final results. Evidently the pressure gradient artifacts associated with hydrostatic consistency or lack of same do not affect the overall results significantly.

Since, in this case, $\eta_t << |(uD)_x|$, $|\omega_\sigma|$, the rigid lid approximation is appropriate as we would have expected from scaling considerations (see Exercise 6.1), though rigid lid models are rare in coastal applications, possibly due to the limitations imposed by steep topography (see Section 4.3). Following the rigid lid formulation, an approximate streamfunction can be calculated for u and ω.

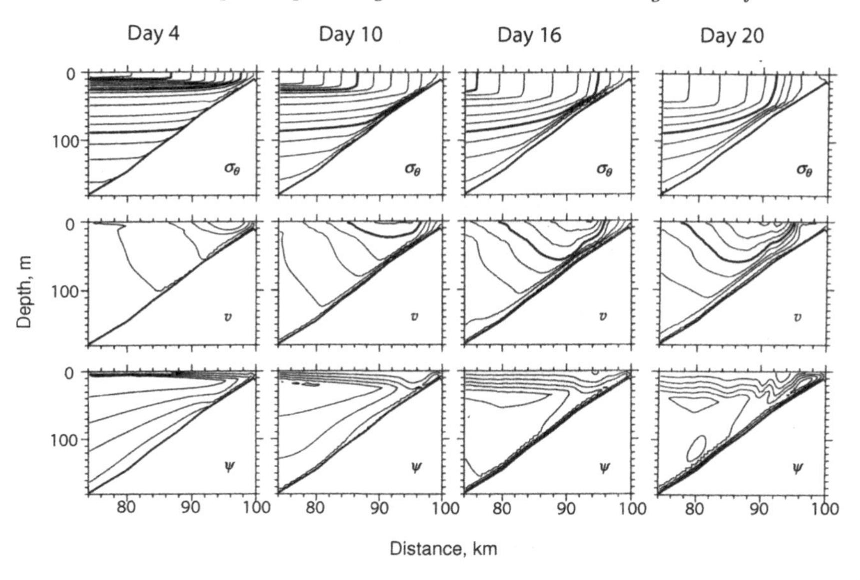

Fig. 6.2 Fields of density σ_θ, alongshore velocity v and streamfunction ψ for the flow in the x–z plane for the basic case at 4, 10, 16 and 20 days. Contour intervals are $0.166\,\mathrm{kg\,m^{-3}}$ for σ_θ, $0.1\,\mathrm{m\,s^{-1}}$ for v and $0.1\,\mathrm{m^2\,s^{-1}}$ for ψ. Bold contours show $\sigma_\theta = 24, 25, 26\,\mathrm{kg\ m^{-3}}$ and $v = -0.5\,\mathrm{m\ s^{-1}}$. Redrawn from Figures 4 and 5 in Allen *et al.* (1995), with permission of the American Meteorological Society.

The fluid is initially at rest, with the density a function of z alone. The overall development of the flow in the basic case is illustrated in Figure 6.2.

Horizontal density gradients appear at the surface by day 4, and the offshore Ekman transport near the surface is balanced by an onshore flow that is initially fairly uniform in depth. Beyond 4 days, at points more than about 15 km offshore, the total offshore transport is roughly equal to the Ekman value $\tau^{(y)}/\rho_0 f$. As the flow develops, the density front associated with the upwelling moves offshore, the layer containing the near-surface offshore flow deepens, and the deep onshore flow is increasingly confined to a narrow layer near the bottom. The alongshore southward coastal jet arises early, and strengthens and moves offshore as the experiment progresses. Eventually, small-scale spatial variability arises near the shore; see especially the streamfunction field inshore of the 60 m isobath at 20 days. At this point the authors report that superinertial temporal variability sets in.

The coastal jet is observed to be geostrophically balanced outside of thin surface and bottom boundary layers, as one might expect from scaling considerations. POM calculates the evolution of the free surface separately from the interior flow, and, since the surface pressure was not retained for analysis, the geostrophic velocity was calculated by integrating the thermal wind equation in the σ-direction from reference values at mid-depth.

Several experiments with different parameterizations of turbulent diffusion were performed, including one with constant eddy diffusivities of heat and momentum. The authors report strong sensitivity to the value of the diffusivities in the constant diffusivity case. They chose equal diffusivities of momentum and heat $= 5.0 \times 10^{-4}\,\mathrm{m}^2\,\mathrm{s}^{-1}$ in the case presented in Allen *et al.* (1995) because those values gave reasonable agreement for the speed of the coastal jet. Beyond the most general schematic picture of coastal upwelling, which is determined by overall balances of mass and momentum, the pictures were quite different. In the case with constant diffusivities, the overall upwelling circulation was weaker than it was in BC, with less dense water near the coast and much gentler slope to the isopycnals inshore within about 8 km of the shoreward boundary. The core of the jet was distinctly weaker, and the onshore flow below the surface layer was not nearly so strongly confined to the interior. Other fairly common parameterizations also showed distinct differences. Most models of the coastal ocean are quite sensitive to the form of the parameterization of diffusivity of heat and momentum, and while a detailed discussion of turbulence is beyond the scope of this book, the coastal modeler must pay close attention to this matter.

An experiment similar to the BC was performed in which the offshore boundary was set at 200 km, and the results were little different. Evidently the artificial boundary at 100 km offshore did no significant damage to the physical model.

Further experiments were performed with finer horizontal resolution. One might be tempted to ascribe the small-scale oscillations in the streamfunction field near the bottom to lack of resolution, perhaps a cell Reynolds number instability, but they arise at about the same wavenumber in a fine-grid experiment ($\Delta x = 0.166\,\mathrm{km}$), in which they would appear to be well resolved. From the published figures it is difficult to say what they are.

Near the bottom and the inshore boundary, the alongshore momentum budget is dominated by an Ekman balance, i.e., a balance between the vertical friction and Coriolis terms. The region where this balance applies spreads offshore over the course of this experiment, from about 12 km offshore at day 10 to about 20 km offshore at day 20. Offshore of this region, no term is as large as either the coriolis or the vertical friction term. A minimum of tendency, probably indicating the southward acceleration of the coastal jet, appears about 12 km offshore on day 10. The tendency reverses sign about 14 km offshore. From there to about 20 km offshore positive tendency and negative advection are the terms with the greatest magnitude. By day 16, the Ekman balance has spread nearly 20 km offshore, where it remains the most prominent feature in the balance.

Along a mid-depth σ-level, on day 10, beyond about 8 km offshore, the positive coriolis term corresponding to onshore flow is balanced by negative tendency, and, to a lesser extent, negative advection, as the southward jet increases in speed. Inshore of 8 km, the σ-level chosen to illustrate the momentum balance approaches the surface and enters a region of offshore, rather than onshore flow. On days 10 and 16 the greatest term is positive vertical friction, balanced by a combination of negative coriolis and advection.

Near the surface, offshore of about 4 km, a balance close to an Ekman balance develops on day 10, with lesser contributions from negative tendency and positive advection terms. Inshore of about 4 km, the large positive vertical friction term is balanced by a combination of coriolis and advection. This general balance continues on day 16, but small-scale variability of tendency and advection become significant at the surface and at mid-depth on day 20. This variability is probably associated with the small scale variability apparent in the streamfunction field shown in the lower right-hand panel of Figure 6.2.

Comparisons of observations with a series of runs of the two-dimensional model were presented by Federiuk and Allen (1995). Observations came from the 1973 Coastal Upwelling Experiment CUE-2, performed off the Oregon shelf; see, e.g., Smith (1981). Data from CUE-2 include hydrographic and current measurements, as well as a line of current meter moorings taken at $45^\circ\,15'$ N, taken during July and August of 1973. In the basic experimental setup, initial conditions were derived from a density field derived from a CTD survey. Initial alongshore velocity was in geostrophic balance with the density field, cross-shore velocity was set to zero and the model was driven with a spatially homogeneous wind field. These conditions were varied systematically in a series of experiments. In one experiment, wind curl was imposed based on observed decorrelation scales. In another experiment, all velocity components were initially set to zero and the initial density was specified as a function of z alone. In all but one case, heat flux was imposed in the form of a spatially uniform diurnal cycle with a daily average of 166 W m^{-2}. One run was performed with zero heat flux.

Over the course of the 60-day experiments the model retains the influence of its initial conditions. In the run with initially flat isopycnals, the jet is weaker and narrower than it is in the basic case. The mean cross-shore velocity components are similar in the two cases, but the advective cooling is stronger in the interior in the case with initially flat isopycnals. This is probably due to the fact that, in the basic case, the mean circulation in the x–z plane is more closely aligned with the isopycnals.

The model reproduces many of the observed features in some form, though it tends to underpredict the cross-shore velocity and its variability, and overpredict the alongshore velocity. These are probably significantly affected by essentially three-dimensional phenomena. In one experiment a negative alongshore pressure gradient (i.e., with pressure decreasing northward) was added, with the result that the mean alongshore velocity and its standard deviation were closer to their observed values, but the pattern in x–z space differed appreciably from those observed.

One can deduce complicated consequences of the interaction of stratified flows with topography from purely two-dimensional conceptual models. Consider, e.g., along-isobath flow over a sloping bottom, in a situation in which the density is initially a function of z alone. Ekman transport in the bottom boundary layer will transport fluid up or down slope, depending on the direction of the current, and result in horizontal density gradients, which will then result in vertical velocity gradients according to the thermal wind relation (MacCready and Rhines, 1993). This thermal shear tends to oppose the flow in the interior of the fluid, with the eventual result that the shear, and hence the viscous stress at the boundary, decays exponentially. MacCready and Rhines' model was formulated in spatial coordinates tangential and normal to a gently sloping bottom. Changes in density resulting from advection within the Ekman layer could be expressed in terms of the product of the vertical component of the up- or down-slope velocity and the buoyancy frequency, so the calculation could be restricted to one space dimension, viz. the vertical, or, more precisely, the normal to the bottom. A two-dimensional model was proposed by Chapman and Lentz (1997) with similar result.

An idealized three-dimensional numerical experiment was performed by Chapman (2000) using the s-coordinate model SPEM (Song and Haidvogel, 1994) in an open channel 600 km long and 400 km wide, with depth increasing linearly in the cross-stream (y) direction from 50 m at $y = 0$ to 400 m at $y = 350$ km, beyond which the bottom was flat. A sponge layer was imposed on the deep side of the channel, i.e., 320 km $\leq y \leq$ 400 km. The numerical model bears out some of the features of the conceptual two-dimensional model, especially in the initial adjustment process, but some bottom stress remains downstream. The conclusion that the bottom stress should actually become arbitrarily small under these circumstances seems counterintuitive (at least to the author), so it is hardly surprising that a three-dimensional experiment showed that some bottom stress remained.

The west coast of North America, with its fairly narrow shelf and relatively straight coastline, should be the place where models do best, in the sense that they should provide a direct insight into physical phenomena in terms of schematic theories. If two-dimensional theories such as those described here are to be successful, they have the best chance here. Still, coastal flow off the west coast of North America is complex, and exhibits essentially three-dimensional features. This is illustrated in the following example.

6.3 Example: A three-dimensional calculation off the coast of California

In this section we describe an application of the Princeton Ocean Model (POM) to the response of the coastal ocean to relaxation of upwelling favorable winds, as described by Gan and Allen (2002). Summer winds off the coast of California are predominantly upwelling favorable, viz. from the north. Models and observations exhibit the familiar upwelling circulation. Alongshore winds, however, are episodic, and, when they weaken, northward currents develop adjacent to the coast, in the absence of south winds. The simple intuitive kinematic explanation is that the north winds drive the southward upwelling jet and pile up water downwind, so we expect a negative pressure gradient (i.e., pressure increasing southward) to develop in response. When the wind relaxes, the remaining pressure gradient drives a current northward. There is, however, no natural southern boundary of this region, so the notion of water somehow piling up downwind must be considered carefully.

This phenomenon of poleward flow developing along with relaxation of upwelling favorable winds is most closely associated with capes, where the jet is observed to separate from the coastline. Some of the coldest water in the region is found inshore of the separated jet.

It turns out that the negative pressure gradients are set up by interactions of the upwelling current system with variations in topography, of which capes are the most prominent. Sharp curvature of the land boundary in the vicinity of capes tends to set up a gradient wind balance (cf. Holton, 1992), i.e., a three-way balance between centrifugal force, pressure gradient and coriolis, as opposed to a geostrophic balance which includes only pressure gradient and coriolis. This serves to accentuate the dynamic decline in pressure on the downstream side of the cape, and thus the negative pressure gradient. This negative pressure gradient also tends to accentuate the onshore flow at depth through a geostrophic balance, which intensifies the upwelling and results in cooler denser water near the coast than would occur in the absence of this pressure gradient.

Gan and Allen (2002) sought to investigate this behavior by performing a process study with POM, implemented with curvilinear coordinates for the California coast. Their domain is shown in Figure 6.3, along with the locations of the mooring lines referred to in the text. Periodic boundary conditions were imposed at the north and south boundaries. The deciding factor for that choice was the importance of alongshore pressure gradients. While periodic boundary conditions cannot support pressure gradients on the scale of the entire domain, it is important to consider the possibility of systematic errors in such pressure gradients that may result from open-boundary conditions, especially given the well-known ill-posedness of the problem, cf. Section 4.11.

The model was initialized with horizontally uniform stratification taken from regional climatology (Levitus and Gelfeld, 1992). The system was spun up with a spatially uniform upwelling-favorable (i.e., southward) wind stress of 0.1 Pa for 10 days. After the initial spinup period, the wind was tapered to zero linearly over a period of 3 days, and then maintained at zero for the remainder of the experiment.

Northward flow develops at the surface inshore of the jet by day 15 as shown in Figure 6.4. Profiles of density and alongshore velocity in the plane defined by the cross-shore direction and depth at station C are shown in Figure 6.5. The development of the poleward flow appears in the lower two left-hand panels. It is evidently associated with the relaxation of upwelling characterized by the change in shape of the isopycnals at distances 25 km from shore and closer. By day 20, the surface water

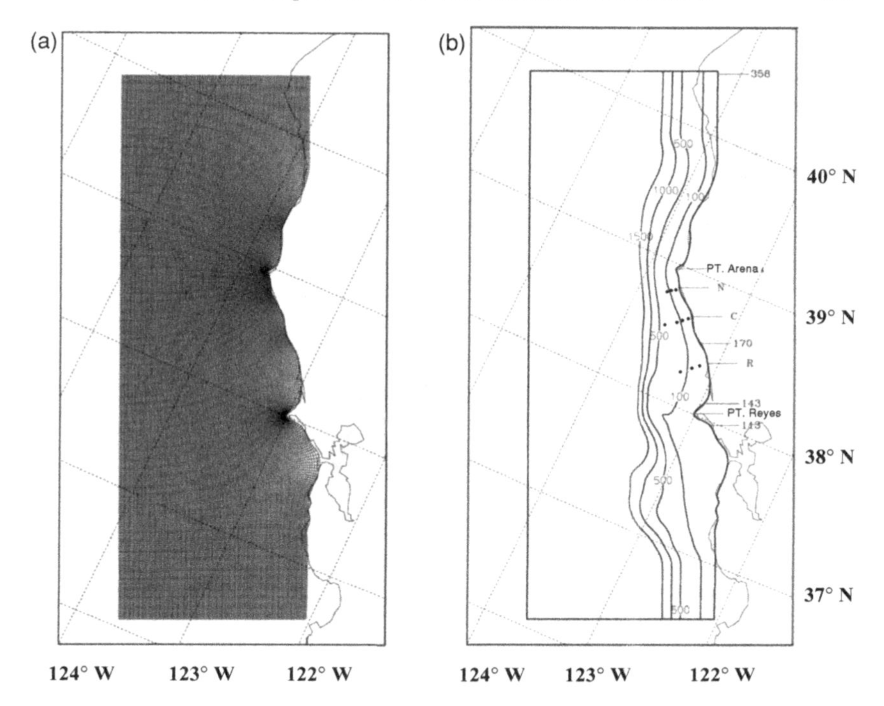

Fig. 6.3 Computational domain for POM study of relaxation of upwelling on the coast of California. Actual geography is shown in light gray; note, e.g., the San Francisco bay near 38° N. (a) Curvilinear computational grid. (b) Topography, showing selected isobaths, as well as the locations of Point Arena, Point Reyes and the mooring lines for the Coastal Ocean Dynamics Experiment (CODE). Mooring locations are shown as filled circles. Note the inclination of the grid from true north and east by the parallels and meridians of latitude and longitude, shown here as thin dotted lines. Redrawn from Figure 3 of Gan and Allen (2002), with permission of the American Geophysical Union.

in this region is nearly uniform in density and no denser than $\sigma_\theta = 25.6$, in marked contrast to the sharper gradients and greater densities that appear on day 10.

At the "N" and "R" lines (see Figure 6.3), the jet moves offshore between days 10 and 15, as it does at the "C" line, but the northward flow inshore of the jet is much weaker, and, to the extent that it exists at all, deeper; see Figure 7 of Gan and Allen (2002). Examination of the term balances in the depth-averaged alongshore momentum equation shows the poleward flow that develops as the wind relaxes and afterward

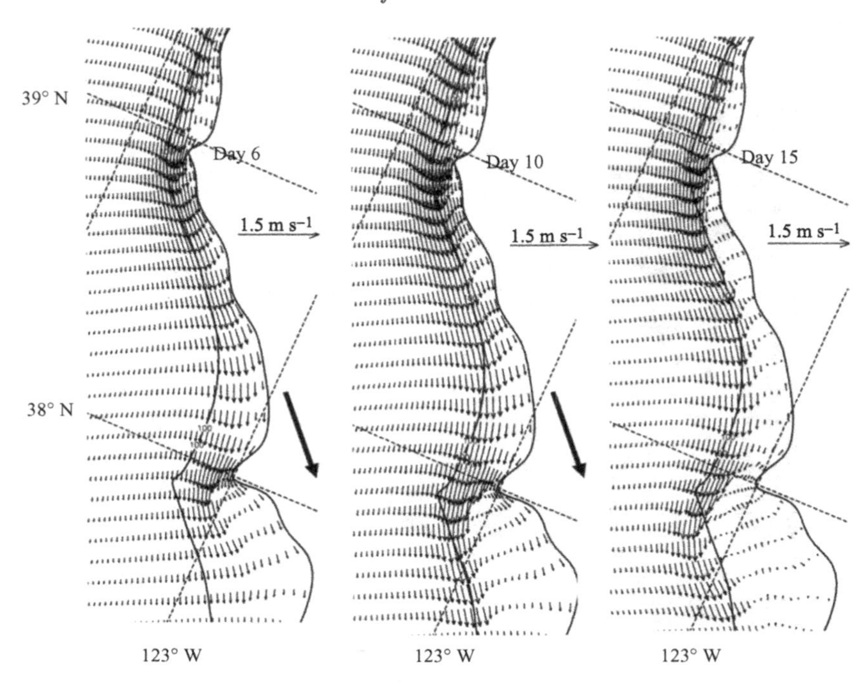

Fig. 6.4 Surface velocity vectors on days 6, 10 and 15. Note the northward flow inshore of the jet north of Point Reyes on day 10, and the recirculation south of Point Reyes. Bold arrows depict the direction of wind stress on days 6 and 10. Redrawn from Figure 5a of Gan and Allen (2002), with permission of the American Geophysical Union.

appears to be associated with a negative pressure gradient (i.e., pressure increasing equatorward).

Figure 6.6 shows the depth-averaged term balances for the cross-shore momentum equation as a function of alongshore distance on day 3 of the experiment at a distance of about 5.5 km from the coast. The alongshore flow over most of the region is geostrophically balanced by day 3, as we would expect from Exercise 6.1. Other terms in the cross-shore momentum balance are negligible away from Point Reyes and Point Arena, where advection becomes important, and the flow is better described by a gradient wind balance. This is reflected in the local minima in the surface pressure at Point Reyes and Point Arena, as shown in the upper panel of Figure 6.6.

A two-dimensional calculation was performed for comparison to the basic case at location 170. Comparison of the development of the

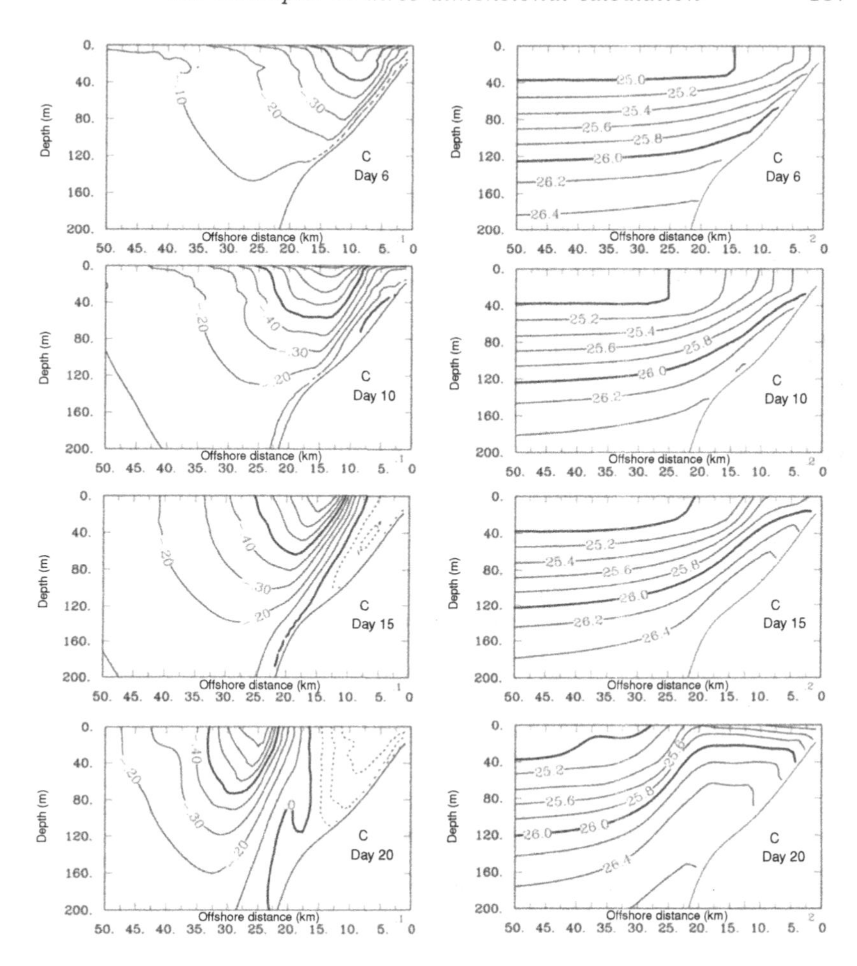

Fig. 6.5 Cross-shore sections of alongshore velocity (left-hand column) and potential density (right-hand column) at the location of the CODE "C" line; see Figure 6.3. Redrawn from Figure 6 of Gan and Allen (2002), with permission of the American Geophysical Union.

alongshore velocity is shown in Figure 6.7. On day 6, the flows are fairly similar. They remain similar by day 10 in that the magnitudes and locations of the peak surface currents are similar, but the jet is slightly stronger and the upwelling front is more strongly sheared in the 2-D case. As the winds relax, the currents in the 2-D calculation remain more strongly sheared, and, by day 20, distinctly stronger. The

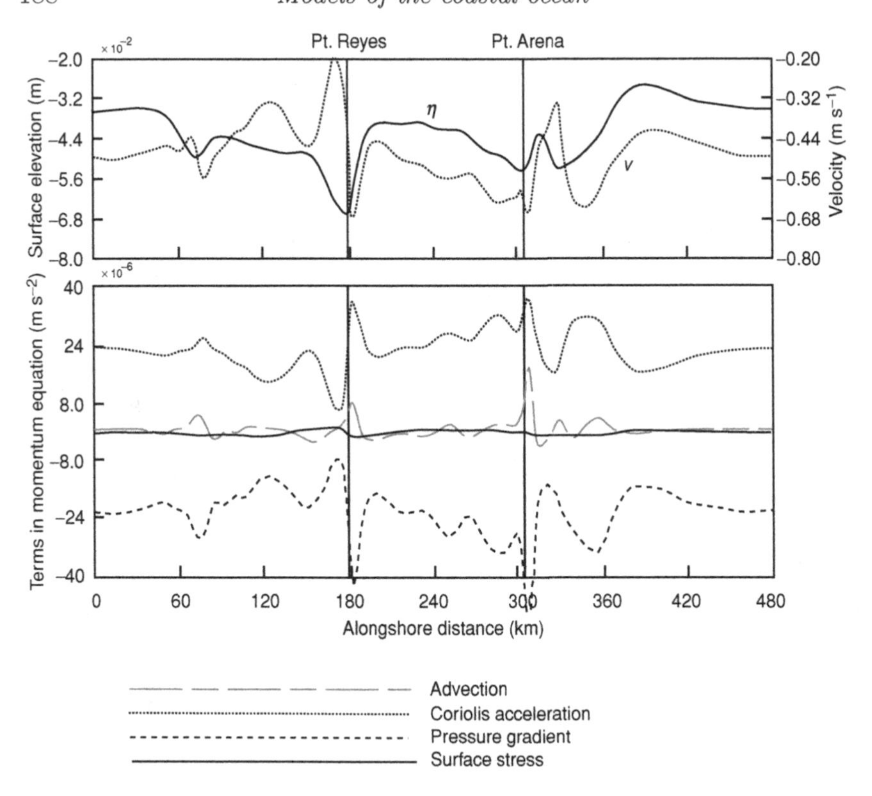

Fig. 6.6 Balance of terms in the depth-averaged cross-shore momentum equation as a function of alongshore distance, about 5.5 km from the coast, averaged over 24 hours on day 3. Redrawn from Figure 12 of Gan and Allen (2002), with permission of the American Geophysical Union.

zero contour appears near the bottom between 10 and 25 km offshore in the three-dimensional calculation. This feature is absent from the two-dimensional calculation. The natural explanation for this is the negative alongshore pressure gradient that tends to decelerate the jet during and after the relaxation of the wind. This is not represented in the 2-D model.

Additional calculations were performed with purely linear dynamics and with a flat bottom. Results are summarized in Figure 6.8. Separation of the jet downstream of Point Reyes is absent in the linear model, so there is no eddy just below Point Reyes, as there is in the nonlinear run. The flow is significantly decelerated adjacent to the coast after the relaxation of the winds – note the panel in the middle row, right-hand

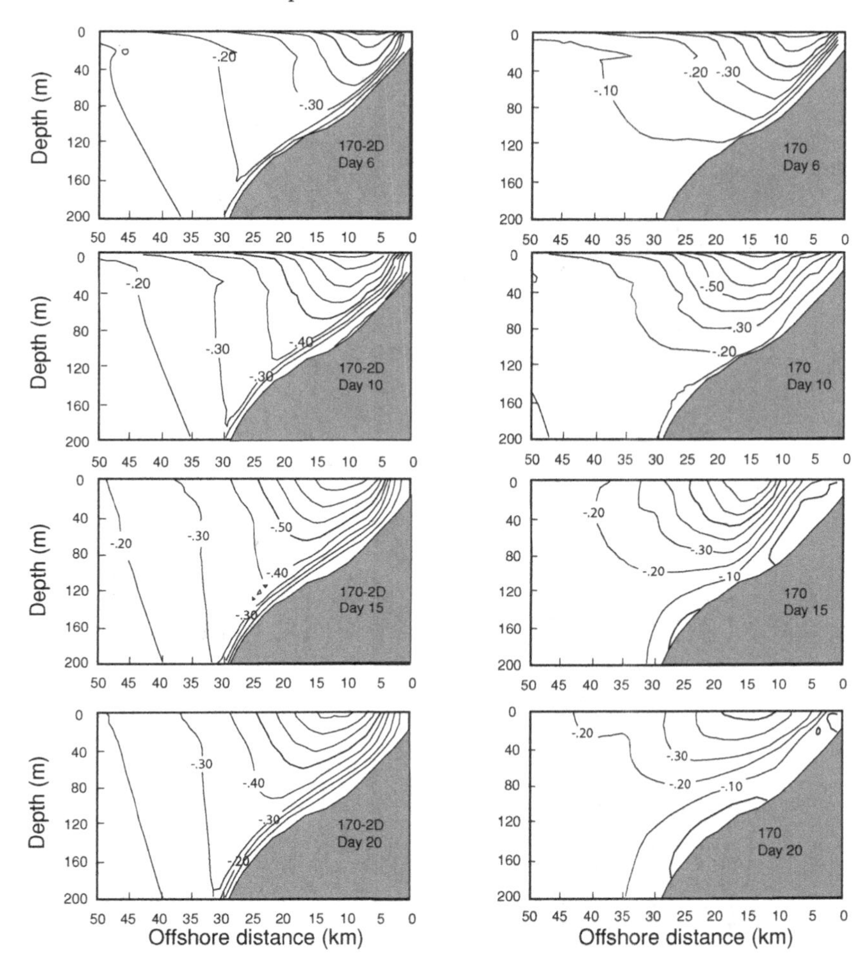

Fig. 6.7 Cross-shore sections of velocity at line 170 (see Figure 6.3) on days 6, 10, 15 and 20. Left-hand panels show results of a two-dimensional calculation. Right-hand panels show results from the basic case. Redrawn from Figure 20 of Gan and Allen (2002), with permission of the American Geophysical Union.

column – but it never reverses, as it does in the basic case. Clearly the linear case cannot reproduce the gradient wind balance.

The flat-bottom case was performed with the coastline as a vertical wall and the shelf at a constant depth of 300 m. In the flat-bottom case, an eddy forms south of Point Reyes, but the effect of relaxation of the wind is much weaker.

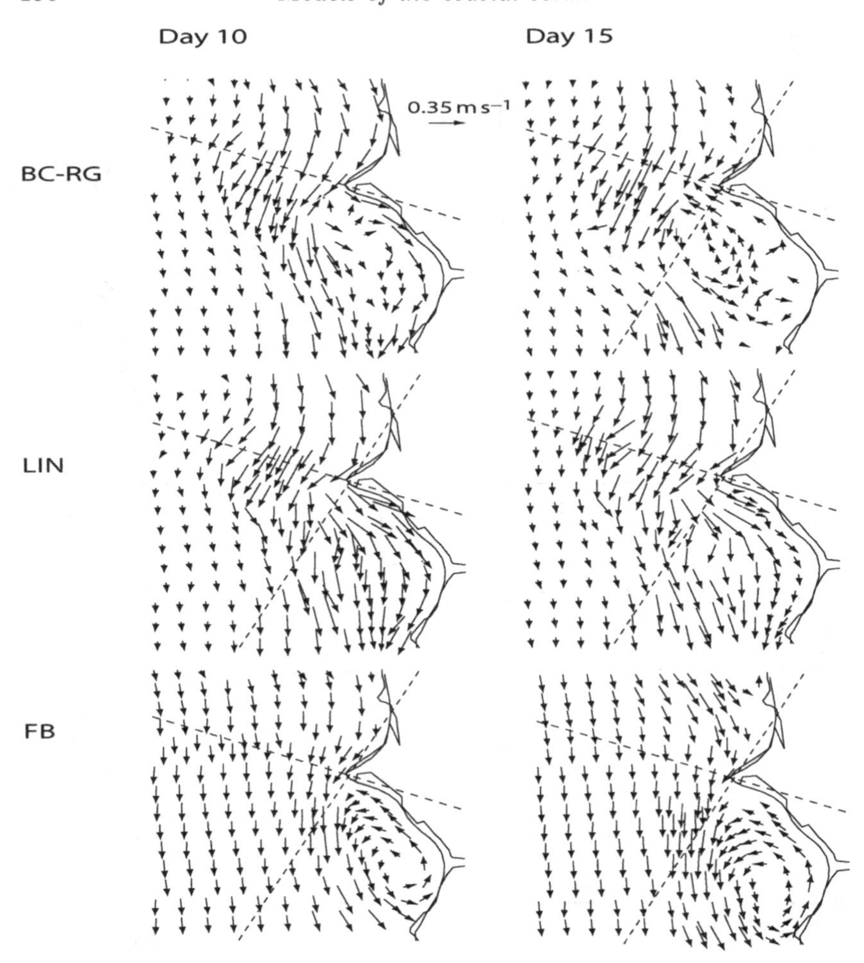

Fig. 6.8 Depth-averaged velocity vectors in the vicinity of Point Reyes on days 10 and 15 for experiments with reduced resolution (top), linear dynamics (center) and flat bottom (bottom). Redrawn from Figure A2 of Gan and Allen (2002), with permission of the American Geophysical Union.

Even what should be a fairly simple coastal flow turns out to be quite complex. The coastal jet is mostly geostrophically balanced, as we would expect from scale analysis, but advection can be important in the presence of sharp topographical features. Two-dimensional calculations can be useful but miss essential features of the flow, as do simulations that neglect nonlinearity and bottom slope.

Clearly, some attempts to infer three-dimensional behavior from two- and even one-dimensional models can be enlightening, but even the most subtle and sophisticated of these can be misleading.

6.4 Example: A finite-element model

Finite-element methods are well suited to models of the coastal ocean because of their ability to represent complex geometry and the ease with which resolution can be varied from one region to another. They can also be quite computationally efficient, in the sense that accurate solutions can be obtained with reasonable demands on computing resources. In this section we discuss a finite-element model of the circulation in the Gulf of Maine, as presented in Lynch *et al.* (1996).

In that work, a finite-element model was used to simulate circulation driven by tidal forcing, wind and heat flux in the Gulf of Maine. They do not present detailed comparisons to observations, but they show that the model reproduces some of the known qualitative features of the circulation and its seasonal variability.

Their model is a fairly simple one mathematically, with triangular elements and C^0 basis functions, i.e., continuous at the boundary but not differentiable. The free-surface elevation is calculated by a shallow-water wave equation that involves second derivatives in time, and to which artificial numerical damping with a coefficient of $2 \times 10^{-4}\,\mathrm{s}^{-1}$ is added, implying a fairly rapid decay rate. Vertical discretization is a fairly complex scheme that follows both the bottom topography and the free surface. Density is interpolated onto horizontal surfaces for computation of baroclinic pressure gradients.

Turbulent exchanges of mass and momentum in the vertical direction are parameterized by the Mellor–Yamada level 2.5 closure scheme (Mellor and Yamada, 1982). Horizontal eddy viscosity is calculated as a function of shear and mesh scale according to a scheme devised by Smagorinsky (1963).

As noted by the authors, this is a natural setting for a finite-element method of this sort. The local value of the internal radius demands resolution no coarser than about 5–10 km and resolution of the frontal circulation that occurs in this region requires meshes of 2 km or so. Equilibration of inflows require that the model cover a spatial extent of the order of four shelf widths, i.e., about 1000 km. Heat and momentum transfers at the surface and bottom require vertical resolutions of the order of 1 m, and the total depth reaches 300 m in the basins and 1000 m

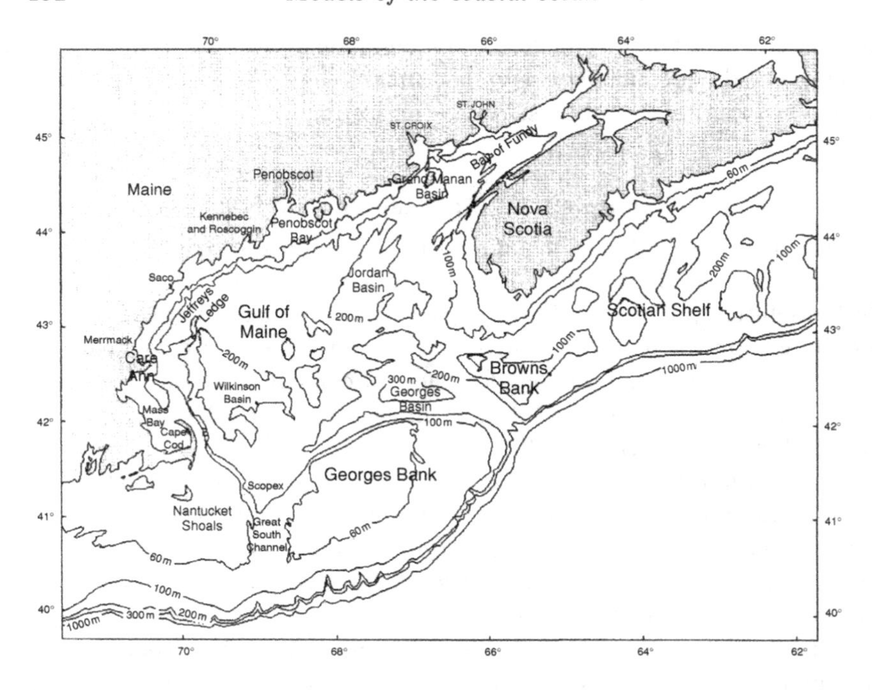

Fig. 6.9 Topography of the Gulf of Maine. Redrawn from Figure 12 of Lynch and Holbroke (1997), with permission of John Wiley and Sons, Ltd.

at the shelf break. The domain is shown in Figure 6.9 and the finite-element mesh is shown in Figure 6.10.

Open-boundary conditions are chosen to enforce total mass conservation and consistency with the internal solution. This is a fairly heavily damped model, and it is probably this heavy damping that allows them to avoid the ill-posedness associated with open-boundary conditions for the inviscid open-boundary problem. Conservation of total mass is clearly an important consideration for coastal situations such as this one that are strongly forced by tides.

The finite-element mesh has 6756 nodes and 12 877 triangular elements. The finest resolution approaches 2 km on the northern flank of Georges Bank. A grid with uniform resolution of 2 km or so would have an order of magnitude more points.

Experiments with M_2 tidal forcing in nonstratified and stratified cases show strong rectification. Vertically averaged currents averaged over a tidal period in the vicinity of Georges Bank show a strong (0.05–0.06 $\mathrm{m\,s^{-1}}$) current along the northern flank in the nonstratified case, as shown in Figure 6.11. In the companion stratified calculation, a

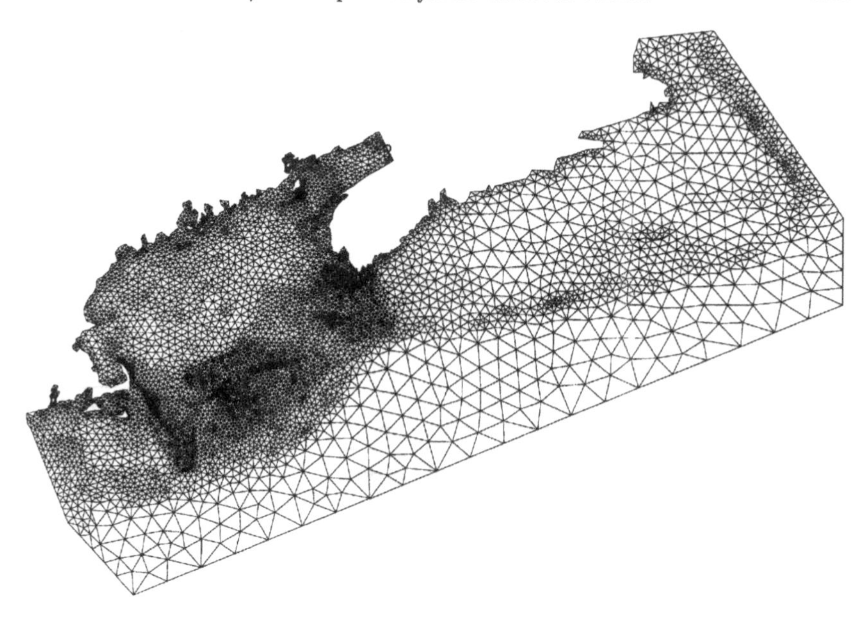

Fig. 6.10 Finite-element mesh for the Gulf of Maine model. Reproduced from Figure 21 of Lynch and Holbroke (1997), with permission of John Wiley and Sons, Ltd.

surface heat flux is imposed that is sufficient to maintain stratification of $3\sigma_t$ in the bank area in order to mimic the summer situation. The stratified result, shown in Figure 6.12 is characterized by stronger currents north of the bank and strengthened recirculation flow in the Great South Channel, evidently in accord with observations of seasonal variation. Earlier work by these authors indicates that much of the increased current strength results from a decrease in viscosity in the stratified case. This fact emphasises the importance of the turbulence model used to calculate the viscosity.

The tidal forcing is present throughout the year regardless of weather conditions, so it is not surprising that rectification of tidal forcing, when it occurs at significant strength, is an important component of the average flow. This is clearly a nonlinear effect. Given the evident importance of nonlinearity and turbulence in the shape of the mean flow, it should not be surprising that complex models are required for accurate simulation of coastal flows. Lynch *et al.* (1996) do not present detailed comparisons to observations, but they show that this full featured model, now the basis of the "Quoddy" model, can simulate the basic features of the complex circulation in the Gulf of Maine, and, by implication, in other places.

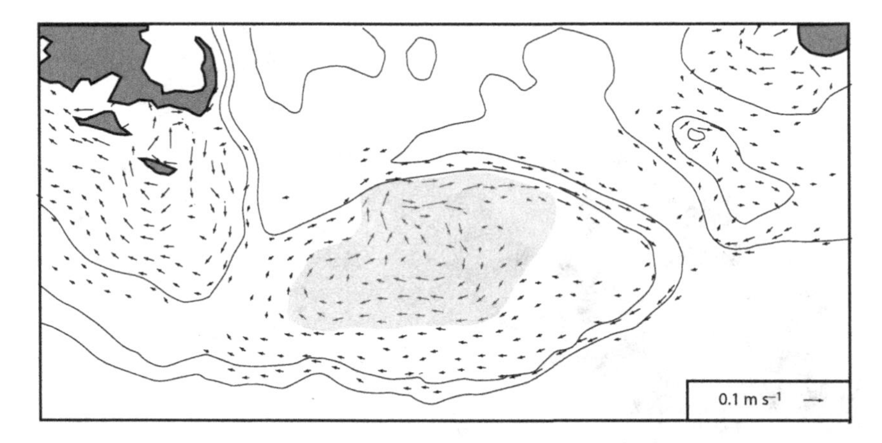

Fig. 6.11 Vertically averaged tidal residual current in the vicinity of Georges Bank, forced with tide only. Cape Cod, Martha's Vineyard and Nantucket are shown in the upper left-hand corner; Cape Sable is shown in the upper right-hand corner. The light gray shaded area near the center is the 60 m contour, indicating Georges Bank. Redrawn from Figure 3 of Lynch *et al.* (1996), with permission of Elsevier.

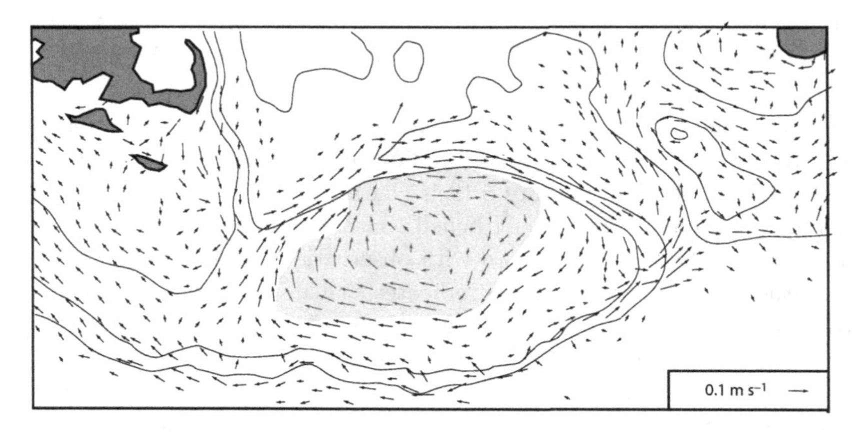

Fig. 6.12 Similar to Figure 6.11 but for the stratified case. Redrawn from Figure 4 of Lynch *et al.* (1996), with permission of Elsevier.

6.5 Summary

Most models of the coastal ocean must deal with detailed topography, forcing by tides and by the atmosphere, and complex dissipation mechanisms. Most coastal models also involve boundary issues. Design of specific models depends on specific modeling objectives and regional differences dictate a wide variety of coastal modeling situations. For this reason coastal models in general have less in common with one another than do models of the open ocean.

Models of the east coast of North America are usually designed with different considerations in mind from those of the west coast. The shelf is broader and the topography more complex. Regions such as the Gulf of Maine cannot be conveniently described, even at a rough approximation, in simple form such as that shown in Figure 6.1. Models of the west coast are often configured to simulate the wind-driven upwelling and downwelling, and associated jets and density structures, that are major determinants of the physical, chemical and biological environments. Wind-driven upwelling and downwelling are present for fundamental physical reasons, and certainly occur in other contexts, but the scale and complex geometry of the northeast coast lead to a different set of modeling requirements, and modeling of other continental margins is carried out with yet different sets of objectives and constraints.

6.6 Exercises

6.1 *The long-wave approximation:* Imagine a schematic continental boundary: the fluid occupies a region defined in horizontal extent by the half-plane $x < 0$ and by $-H(x) < z < 0$ in the vertical. The scale of the assumed motion is much longer in the y ("alongshore") direction than in the x ("cross-shore") direction. Let the alongshore length scale be L, the cross-shore length scale be l and the vertical scale be D.

Write the linearized primitive equations:

$$u_t + p_x - fv = 0, \tag{E6.1}$$

$$v_t + p_y - fu = 0, \tag{E6.2}$$

$$u_x + v_y + w_z = 0, \tag{E6.3}$$

$$\rho = \rho_0[1 + \bar{\rho}(z) + \rho_1(x, y, z, t)], \tag{E6.4}$$

$$p_z + g\rho_1 = 0, \tag{E6.5}$$

$$\rho_{1t} + w\rho_1' = 0. \tag{E6.6}$$

Let the alongshore velocity scale be V. Balancing terms in (E6.3) leads us to scale w by $(D/L)V$ and u by $(l/L)V$. We are interested in subinertial motions, i.e., motions on a timescale σ^{-1} such that $\sigma << f$.

(a) With the given scales, find the appropriate scale for pressure.

(b) Set $(\sigma/f)(L/l) = 1$ and derive the *long-wave equations*:

$$p_x - v = 0, \qquad \text{(E6.7)}$$

$$v_t + p_y + u = 0, \qquad \text{(E6.8)}$$

$$u_x + v_y + w_z = 0, \qquad \text{(E6.9)}$$

$$p_z + \rho_1 = 0, \qquad \text{(E6.10)}$$

$$\rho_{1t} - Sw = 0, \qquad \text{(E6.11)}$$

$$S = \frac{N^2}{f^2}\left(\frac{D^2}{L^2}\right); \quad N^2 = -g\bar{\rho}'. \qquad \text{(E6.12)}$$

(c) Derive an equation for conservation of energy from (E6.7)–(E6.12).

(d) Show, from scaling of the free-surface boundary condition, that the rigid lid approximation is a good one.

6.2 *Barotropic waves in the coastal environment*

(a) Derive an expression for potential vorticity by eliminating the pressure terms in (E6.7) and (E6.8) and substituting from the equation of continuity (E6.9).

(b) Integrate your potential vorticity equation from the scaled bottom $z = -H_0/D$ to the free surface $z = \eta$. Apply the linearized free-surface boundary condition at the top and the condition of no normal flow at the bottom to arrive at the single equation for the pressure p:

$$\left[p_{xx} + \frac{H_{0x}}{H_0}p_x - F\frac{D}{H_0}p\right]_t + \frac{H_{0x}}{H_0}p_y = 0, \qquad \text{(E6.13)}$$

where $F = l^2 f^2/gD$.

(c) Consider first the case of a flat bottom, i.e., $H_0 = D$, and look for solutions of the form $p = \phi(x)\exp[\mathrm{i}(ky - \omega t)]$. Find the speed of these waves and show that:

(i) they are non-dispersive;

(ii) they travel with the coast on their right, relative to an observer looking in the direction of propagation;

(iii) the alongshore velocity is geostrophically balanced and the cross-shore velocity vanishes.

(d) Now consider idealized topography in which the continental shelf has uniform slope out to a distance l from the coast, and then drops sharply down to a depth D beyond, i.e.,

$$H_0(x) = \begin{cases} D, & x < -l, \\ \gamma D x/l, & -l < x < 0. \end{cases} \tag{E6.14}$$

Again, seek traveling waves, i.e., solutions of the form $p = \phi(x)\exp[\mathrm{i}(ky - \omega t)]$; we find $\phi = A\exp(F^{1/2}x)$ for some constant A outside the continental shelf, i.e., for $x < -l$. On the shelf, i.e., $-l < x < 0$, ϕ satisfies

$$x\frac{d^2\phi}{dx^2} + \frac{d\phi}{dx} - \phi\left(\frac{k}{\omega} - \frac{F}{\gamma}\right) = 0. \tag{E6.15}$$

Transform (E6.15) into Bessel's equation of order zero, and derive the dispersion relation for these waves by matching the solution on the shelf to the solution beyond the shelf.

(e) In the case of a vertical wall at the boundary, there is only one family of waves, i.e., that given by $p \propto \exp(F^{1/2}x)$. In the case of a uniformly sloping shelf, there is a family of waves corresponding to the repeated roots of the zero-order Bessel function. This is apparently true, independent of l, i.e., for an arbitrarily narrow shelf. Why? Where did these waves come from?

6.3 *Waves in the stratified coastal environment*: Now consider the stratified case for (E6.7)–(E6.12).

(a) Derive a single partial differential equation for the pressure p from (E6.7)–(E6.12).

(b) Seek waves propagating in the alongshore direction, i.e., $p = F(x,t)\exp(\mathrm{i}[ky - \omega t])$ and write the appropriate boundary conditions.

(c) Use the expression for energy you found in Exercise 6.1(c) to define a bilinear form, and derive an orthogonality condition for your wave solutions relative to this form.

(d) How would you calculate the wave solutions and their speeds numerically?

7
Models of the tropical ocean

7.1 Introduction

Some of the most notable early successes in ocean modeling came in the 1980s in models of the tropical oceans, with most being applied to the tropical Pacific Ocean due to its role in interannual climate variability. The success of these models, many of which involved coarsely resolved simple linear dynamics, was due to a variety of factors. The standard by which success of these and many other models were judged was by comparison to pressure anomaly data in the form of sea level height anomalies or dynamic height increments. Sea level data carry the El Niño–Southern Oscillation (ENSO) signal, which is the phenomenon of greatest interest in the study of the tropical Pacific. Relatively rapid progress in models of the tropical ocean, as opposed to the mid-latitude or polar oceans is due, at least in part, to the agenda in tropical oceanography, which is focused on climate variability.

The success of the simple models stems from the fact that to some extent they contain the basic phenomenology of the interaction of large-scale low-frequency atmospheric variability with the large-scale low-frequency behavior of the fluid. When the early studies were performed, the only wind data sets available were coarsely resolved in space and time, and the only available ocean data sets with long enough time series to capture interannual climate variability consisted of depth-integrated pressure anomalies, i.e., data from tide gauges and expendable bathythermograph (XBT) data from volunteer observing ships (VOS); this latter data source is usually processed as dynamic height. Studies such as that of Rebert *et al.* (1985) established, through detailed analysis of observed data, the relation between sea level, thermocline depth, heat content and dynamic height relative to 500 m in the Pacific, confirming, at least in the Pacific Ocean, the applicability of the theorist's reduced-gravity ocean model, with a manageable number of active layers separated from the deep ocean by a sharp thermocline.

These simple models, effective as they are in simulating the large scale low frequency response of the tropical ocean, have shortcomings, which have become more apparent as the community has gained experience with coupled models of the ocean and atmosphere, and as more data have become available. Useful sea surface temperature (SST) maps for study of seasonal to interannual variability in the tropical ocean have been available over several ENSO cycles (see, e.g., Reynolds and Smith, 1994), and there is indication that further improvements in SST data sets may be available in the near future (see, e.g., Wentz *et al.*, 2000). Multiple wind data sets are now available, many more than can be conveniently listed here, including long time series of monthly mean winds over the tropical Pacific compiled by Florida State University (FSU; see Stricherz *et al.*, 1992) and reanalyses generated by weather services. Other wind data sets based on remote sensing are available (e.g., Liu *et al.*, 1998; Atlas *et al.*, 1996). A study of the effect of sampling error in scatterometer data sets appeared in the paper by Schlax *et al.* (2001), and the effect of temporal and spatial smoothing of wind data on a simulation of the tropical Pacific was presented by Chen *et al.* (1999). Perhaps most importantly, the wealth of data collected as part of the ten year Tropical Oceans and Global Atmospheres (TOGA) program is now available to the community and the network of buoys from the TOGA Atmosphere-Ocean (TAO) array in the Pacific remains in place long after the first moorings were deployed. The first moored instruments of the Pilot Research Array for the Tropical Atlantic (PIRATA), modeled in part on TAO, have been deployed, and a similar array of moored instruments is planned for the Indian Ocean. One obvious implication of this wealth of data in the tropical ocean is that the weaknesses of the simplified models that performed so well with available data two decades ago have become more apparent and more readily quantified.

For phenomena that depend on the details of the vertical structure of the ocean, more complex models are needed. One important example is the exchange of heat with the atmosphere. While air–sea exchanges are beyond the scope of this text, any model intended to be used to simulate the coupled behavior of the ocean and the atmosphere, or simply simulate the evolution of SST, must produce estimates of the temperature of the water entrained at the base of the mixed layer. Any model that makes the obviously faulty assumption of uniform vertical structure throughout the ocean basin can be expected to run into trouble in this case. Another problem is that the simple models tend to have highly idealized boundaries. While the effects of boundary conditions that are only roughly in agreement with the elementary facts of geography are not precisely known, few doubt that the crude treatment of boundaries in simplified models makes for difficulty in interpreting the results in terms of the real world (see, e.g., Perez *et al.*, 2005).

One distinguishing feature of the dynamics of the tropical ocean is the appearance of a family of zonally propagating waves with meridional structure that decays rapidly away from the equator. Much of the seasonal to interannual variability of the equatorial oceans can be described in terms of these easily characterized waves; in fact, in a narrow band of latitudes surrounding the equator, comparisons of the output of complex GCMs with observed dynamic height anomalies show little improvement over the simplest linear models based on wave dynamics. We begin this chapter with a brief description of these waves.

7.2 Waves in the equatorial ocean

The forms of waves in the equatorial ocean can be derived from the linearized primitive equations in a form similar to the f-plane linearized primitive equations (4.53)–(4.57). In the neighborhood of the equator, we write the coriolis term $f = \beta y$, recognizing the fact that the coriolis acceleration vanishes at the equator. This distinguishes the equatorial β-plane from the f-plane or the mid-latitude β-plane. As in the cases of the f-plane or mid-latitude β-plane, the result of separating vertical from horizontal coordinates on the equatorial β-plane is an equation of motion for each mode that is formally identical to the shallow-water system, with speed c derived from the eigenvalue problem (4.64). In our case, the equations for each vertical mode become

$$U_t + P_x - \beta y V = 0,$$
$$V_t + P_y + \beta y U = 0,$$
$$P_t + c^2(U_x + V_y) = 0.$$

If we now rescale U and V by c, P by c^2, t by $(c\beta)^{-1/2}$ and x and y by $(c/\beta)^{1/2}$ the equations become

$$U_t + P_x - yV = 0, \tag{7.1}$$
$$V_t + P_y + yU = 0, \tag{7.2}$$
$$P_t + U_x + V_y = 0. \tag{7.3}$$

The quantity $(c/\beta)^{1/2}$ is often called the *equatorial deformation radius.* As in the earlier case of the shallow-water equations, solutions to these homogeneous equations can be written as linear combinations of free waves. In this special case, the disturbances associated with these waves decay strongly away from the equator. The simplest illustration of the phenomenon of equatorially trapped wave motion can be found by seeking an analog of simple shallow-water waves in a nonrotating fluid, perhaps motivated by the fact that f vanishes at the equator. Shallow-water waves in a nonrotating frame are longitudinal waves, i.e., the component of velocity transverse to the direction of propagation is zero. We therefore look for zonally propagating waves with V set equal to zero in (7.1)–(7.3). This leads to the ordinary one-dimensional shallow-water equations for U and P:

$$U_t + P_x = 0, \tag{7.4}$$

$$P_y + yU = 0, \tag{7.5}$$

$$P_t + U_x = 0. \tag{7.6}$$

With this scaling, the waves travel at unit speed, and the solutions must be of the form

$$U(x, y, t) = A_+(y)f(x + t) + A_-(y)g(x - t), \tag{7.7}$$

$$P(x, y, t) = -A_+(y)f(x + t) + A_-(y)g(x - t). \tag{7.8}$$

The forms of A_+ and A_- are determined by (7.5), which leads to

$$-A'_+ + yA_+ = 0, \tag{7.9}$$

$$A'_- + yA_- = 0, \tag{7.10}$$

so we must have

$$A_+ \propto \mathrm{e}^{y^2/2}, \tag{7.11}$$

$$A_- \propto \mathrm{e}^{-y^2/2}. \tag{7.12}$$

Equation (7.11) is not a physically realizable solution on the infinite equatorial β-plane, so we are left with the conclusion that the only waves we will observe of this family are the nondispersive eastward propagating waves, with motion decaying strongly away from the equator, characterized by (7.12). These are the *equatorial Kelvin waves.* They resemble the well-known coastal Kelvin waves in that the velocity is in the direction of propagation, they only propagate in one direction, the velocity and the pressure gradient in the transverse direction to the wave propagation are geostrophically balanced according to (7.5), and the motion is confined to a region near the equator, as velocity and pressure anomalies associated with coastal Kelvin waves are confined to a region near the coast.

The Kelvin waves are one member of a larger family of of equatorially trapped wave solutions to (7.1)–(7.3). Because of the existence of this family of equatorially trapped solutions to (7.1)–(7.3), the narrow band of latitudes where these disturbances are significant is sometimes known as the "equatorial waveguide."

To find other trapped wave solutions, first take the curl of the momentum equations (7.1)–(7.2) to find

$$\zeta_t + y(U_x + V_y) + V = (\zeta - yP)_t + V = 0, \tag{7.13}$$

where $\zeta = V_x - U_y$. U and P can be eliminated to form a single equation in V. Begin by eliminating U between the momentum equations to form

$$V_{tt} + P_{yt} + y^2 V - yP_x = 0. \tag{7.14}$$

Next, differentiate the mass equation (7.3) twice:

$$P_{ytt} + (U_x + V_y)_{yt} = 0. \tag{7.15}$$

Differentiate (7.14) with respect to time and combine the result with the above expression to find

$$V_{ttt} - (U_x + V_y)_{yt} + y^2 V_t - yP_{xt} = 0. \tag{7.16}$$

Now differentiate (7.13) with respect to x and use the result to eliminate the terms in U and P, to form

$$\left[V_{tt} + y^2 V - (V_{xx} + V_{yy})\right]_t - V_x = 0. \tag{7.17}$$

Now look for wavelike solutions, i.e., solutions of the form $V = A(y)\mathrm{e}^{\mathrm{i}(kx-\omega t)}$. Substitution into (7.17) leads to

$$A'' - y^2 A + \left(\omega^2 - k^2 - \frac{k}{\omega}\right) A = 0. \tag{7.18}$$

This is an eigenvalue problem for the differential operator $\mathrm{d}^2/\mathrm{d}y^2 - y^2$. Its eigenfunctions are the *Hermite functions*

$$\psi_n = H_n(y) \frac{\mathrm{e}^{-y^2/2}}{(2^n n! \pi^{1/2})^{1/2}}, \tag{7.19}$$

where $H_n(y)$ is the *Hermite polynomial* of degree n. In order for ψ_n to be an eigenfunction of $\mathrm{d}^2/\mathrm{d}y^2 - y^2$, we must have

$$H_n'' - 2yH_n' - H_n = \lambda_n H_n, \tag{7.20}$$

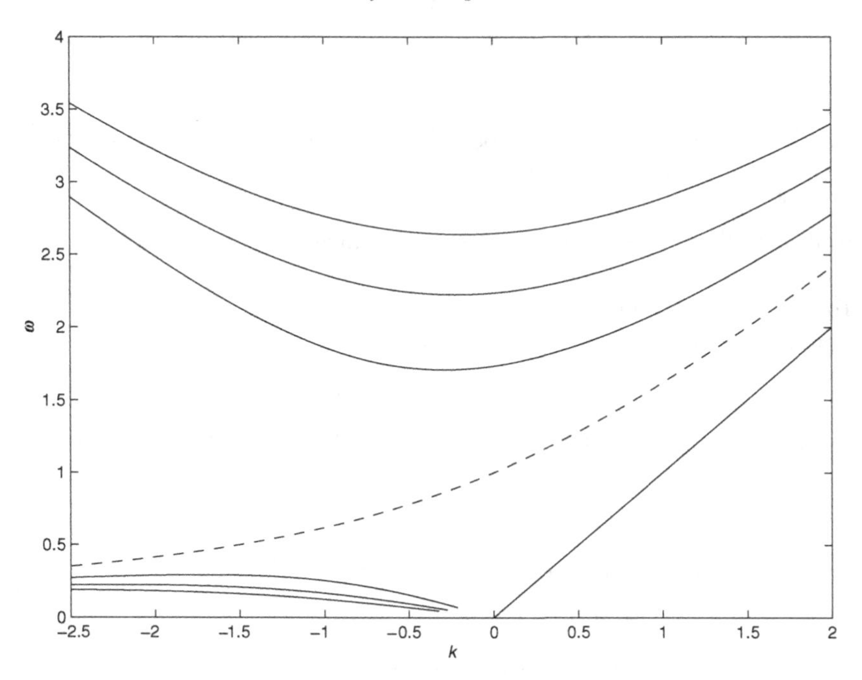

Fig. 7.1 Dispersion curves for equatorially trapped waves, i.e., solutions of (7.22). The dashed curve depicts the dispersion relation for the Yanai wave, corresponding to $n = 0$ in (7.22). Above the dashed curve are the gravity waves corresponding to meridional modes $n = 1, 2, 3$, increasing vertically, so for a fixed zonal wavenumber, higher meridional modes have higher frequencies. Curves in the lower left-hand quadrant below the dashed curve correspond to the equatorial Rossby waves. Higher meridional mode Rossby waves have lower frequencies at fixed wavenumber, and so propagate more slowly. The straight line in the lower right-hand quadrant corresponds to the equatorial Kelvin waves.

for some scalar λ_n. The Hermite polynomials H_n satisfy the differential equation (see, e.g., Copson, 1962)

$$H_n'' - 2yH_n' + 2nH_n = 0, \tag{7.21}$$

so we must have $\lambda_n = 2n + 1$. Our family of waves therefore has the dispersion relation

$$\omega^2 - k^2 - \frac{k}{\omega} = 2n + 1. \tag{7.22}$$

Solutions of (7.22) are plotted in Figure 7.1.

From the figure we see that there are a number of different families of waves. For $n \geq 1$ there are two families of waves, the gravity waves and the Rossby waves. The highest-frequency waves are the gravity waves; the dispersion curves for the gravity waves for $n = 1, 2, 3$ are the three uppermost curves. For fixed zonal wavenumber k, the gravity waves increase in frequency with increasing n, so the topmost curve in the figure is the dispersion curve for the gravity wave with $n = 3$. The dispersion curves for the Rossby waves are the three solid curves in the lower left-hand quadrant. As in the mid-latitude β-plane, the Rossby waves have negative phase velocities. Group velocities of Rossby waves can take both signs, with the lowest wavenumber waves having westward group velocities. Rossby waves decrease in frequency with increasing n, so the lowermost curve in the lower right-hand quadrant represents the $n = 3$ Rossby mode. The nondispersive Kelvin wave appears as the straight line with unit slope in the lower right-hand quadrant. The Kelvin wave is the solution to (7.22) for $n = -1$. The dashed line represents the physically meaningful solution to (7.22) with $n = 0$. This wave approaches the Kelvin wave for large positive zonal wavenumbers and the Rossby wave for large negative wavenumbers. It is referred to as the mixed Rossby-gravity wave, or Yanai wave.

From our solution for V we may reconstruct U and P by writing

$$q = P + U, \tag{7.23}$$

$$r = P - U. \tag{7.24}$$

From (7.1) and (7.3) we find

$$q_t + q_x + V_y - yV = 0, \tag{7.25}$$

$$r_t - r_x + V_y + yV = 0. \tag{7.26}$$

If we write

$$V = V_n \psi_n(y) \mathrm{e}^{\mathrm{i}(kx - \omega t)}, \tag{7.27}$$

we find, by differentiating (7.19) and using the identities

$$H_n' = 2nH_{n-1}, \tag{7.28}$$

$$H_{n+1} - 2yH_n + 2nH_{n-1} = 0, \tag{7.29}$$

that

$$V_y - yV = -2^{1/2}(n+1)^{1/2}\psi_{n+1} V_n \mathrm{e}^{\mathrm{i}(kx-\omega t)}, \tag{7.30}$$

$$V_y + yV = 2^{1/2} n^{1/2} \psi_{n-1} V_n \mathrm{e}^{\mathrm{i}(kx-\omega t)}. \tag{7.31}$$

We may then write $q = Q_n \psi_{n+1} e^{i(kx-\omega t)}$ and $r = R_n \psi_{n-1} e^{i(kx-\omega t)}$ and solve (7.25) and (7.26) to find

$$Q_n = \frac{i 2^{1/2} (n+1)^{1/2} V_n}{\omega - k}, \tag{7.32}$$

$$R_n = \frac{i 2^{1/2} n^{1/2} V_n}{\omega + k}. \tag{7.33}$$

Finally,

$$U = \frac{1}{2}(q - r) = \frac{i V_n}{2^{1/2}} \left(\frac{(n+1)^{1/2}}{\omega - k} \psi_{n+1} + \frac{n^{1/2}}{\omega + k} \psi_{n-1} \right) e^{i(kx-\omega t)}, \tag{7.34}$$

$$P = \frac{1}{2}(q + r) = \frac{i V_n}{2^{1/2}} \left(\frac{(n+1)^{1/2}}{\omega - k} \psi_{n+1} - \frac{n^{1/2}}{\omega + k} \psi_{n-1} \right) e^{i(kx-\omega t)}. \tag{7.35}$$

7.3 Simple models of the tropical oceans

The first successful models of the tropical oceans (e.g., Busalacchi *et al.*, 1980, 1981; Cane and Patton, 1984) were linear models that used finite-difference methods to solve the primitive equations on the equatorial β-plane. Most, but not all of these models (an exception may be found in Smedstad and O'Brien (1991)) are based on the separation of variables recipe, and they work by solving (7.1)–(7.3) for each of a predetermined number of vertical modes. This necessarily implies that such models contain the assumption that the wave speed c_m corresponding to the mth vertical mode is constant for each m, despite the fact that the thermocline in the tropical Pacific shoals to the east to the extent that the depth of the thermocline off the coast of Peru is only half that off New Guinea.

Based on simple reduced gravity considerations ((3.18)–(3.19)) one would expect a variation in the wave speeds of 40% over the Pacific basin. Wave speeds are typically calculated from vertical profiles taken around 160° W (see, e.g., Cane, 1984), so they are representative of most of the interior of the basin. Waves do not propagate undisturbed across the entire basin; rather, they are forced by the wind, which itself varies on large scales. If we figure that the error in the wave speed is 20% and the wave propagates over half the basin, then we would expect the phase error of the model Kelvin wave in the Pacific to be perhaps 10 days, figuring 3 months to cross the entire basin. In fact, the wave speeds vary rather less than that, due to the nature of the stratification in the Pacific basin.

Models used to analyze large scale variability in the tropical oceans must include forcing by wind and possibly heat flux, so they must solve inhomogeneous equations of the form

$$U_t + P_x - yV = \tau^{(x)}, \tag{7.36}$$

$$V_t + P_y + yU = \tau^{(y)}, \tag{7.37}$$

$$P_t + U_x + V_y = 0. \tag{7.38}$$

In this modal decomposition, the stress components $\tau^{(x)}$ and $\tau^{(y)}$ must be projected onto the vertical modes. This is often facilitated by a convenient assumption that the surface stress is spread uniformly over a well mixed layer, typically 50 m deep; see Cane (1984).

Typical values of c for the tropical ocean calculated from the eigenvalue problem (4.67) are of the order of $3\,\mathrm{m\,s^{-1}}$ for the first baroclinic mode. This means that for $\beta = 2.4 \times 10^{-11}\,\mathrm{m^{-1}\,s^{-1}}$ we find the typical time scale for the first baroclinic model to be about 1.4 days, and the equatorial deformation radius for this mode to be about 360 km. Observed wave speeds are distinctly slower. Equations (7.36)–(7.38) were derived under the assumption that the motions involved were small disturbances of a fluid at rest. Much of the reduction in speed and deviation from the idealized Hermite function profiles can be explained by relaxing this assumption. Background meridional shears alter the meridional potential vorticity gradient, and interaction of the motions with the background currents affect the wave speeds, cf. Chelton *et al.* (2003), Perez *et al.* (2005).

The long-wave approximation

The assumption that the timescales and zonal space scales of interest for climate studies are much longer than $(\beta c)^{-1/2}$ and $(c/\beta)^{1/2}$ respectively leads to further simplification. If we now rescale (7.36)–(7.38) according to a new dimensional scale L for the zonal coordinate x and a new dimensional timescale T so that $(\beta c)^{-1/2}/T \approx (c/\beta)^{1/2}/L = \epsilon << 1$, derivatives with respect to x and t become $O(\epsilon)$ quantities, and dimensional consistency in (7.38) requires that $v = \epsilon u$. Equations (7.39)–(7.41) become

$$U_t + P_x - yV = \tau^{(x)}, \tag{7.39}$$

$$P_y + yU = \tau^{(y)}, \tag{7.40}$$

$$P_t + U_x + V_y = 0. \tag{7.41}$$

In writing (7.39)–(7.41) we have arbitrarily kept the stress terms $\tau^{(x)}$ and $\tau^{(y)}$. The reader should note that the left-hand side of (7.39) is uniformly $O(\epsilon)$ while the left-hand side of (7.40) is uniformly $O(1)$, so if $\tau^{(x)}$ and $\tau^{(y)}$ are the same magnitude, $\tau^{(y)}$ will be $O(\epsilon)$ with respect to the left-hand side of the equation. This reduction of (7.1)–(7.3) to (7.39)–(7.41) is called the *long-wave approximation.*

Wavelike solutions to the homogeneous form of (7.39)–(7.41) can be obtained by following the derivation beginning with (7.13). The dispersion relation for the resulting waves is the same as that which follows from simply discarding quadratic terms in k and ω in (7.22). This leads to

$$\omega = -\frac{k}{2n+1}. \tag{7.42}$$

In this approximation, the waves are not dispersive. The gravity waves do not appear at all. Only the Kelvin wave (corresponding to $n = -1$) and the long Rossby waves remain. The solution $\omega = -k$ corresponding to $n = 0$ is not physically realizable, so this approximation also does not include the Yanai wave.

Advantages and limitations of explicit wave models

We may include the forcing terms in this description of waves on the equatorial β-plane by expanding the zonal component of the forcing in terms of Hermite functions, i.e.,

$$\tau^{(x)} = \sum_{0}^{N} \hat{\tau}_n(x,t)\psi_n(y), \tag{7.43}$$

and neglecting the meridional component of forcing, since, as noted above, if the zonal and meridional winds are the same order of magnitude, then the forcing term in (7.40) will be higher order than the other terms.

A simple model of the tropical ocean can be constructed directly in terms of the nondispersive Kelvin and Rossby waves derived under the long wave approximation (see, e.g., Cane and Sarachik, 1981). The velocity and pressure components can be written as an expansion in wave modes of the form

$$\begin{pmatrix} U \\ P \end{pmatrix} = \frac{a_k}{2^{1/2}} \mathrm{e}^{-y^2/2} \begin{pmatrix} 1 \\ 1 \end{pmatrix} + \sum_{j=1}^{M} \frac{r_j}{2 \cdot 2^{1/2}} \mathrm{e}^{-y^2/2} \begin{pmatrix} (j+1)^{-1/2} H_{j+1}(y) - j^{-1/2} H_{j-1}(y) \\ (j+1)^{-1/2} H_{j+1}(y) + j^{-1/2} H_{j-1}(y) \end{pmatrix}. \tag{7.44}$$

a_k and r_j are the Kelvin and Rossby mode amplitudes respectively, and their amplitudes evolve in time according to the simple advection equations

$$\frac{\partial a_k}{\partial t} + \frac{\partial a_k}{\partial x} = \tau_k(x, t), \tag{7.45}$$

$$\frac{\partial r_j}{\partial t} - \frac{1}{2j+1} \frac{\partial r_j}{\partial x} = \tau_j(x, t). \tag{7.46}$$

where τ_k is the Kelvin wave forcing and τ_j is the jth mode Rossby wave forcing (see Exercise 7.1).

Boundary conditions at the eastern and western boundaries for this expansion are approximate expressions of the condition that there be no mass flux into the boundary. These conditions are derived from (7.44). At the eastern boundary $x = x_{\mathrm{E}}$, the condition that the zonal velocity component u vanishes becomes a relation between the wave amplitudes that can be derived by choosing the wave amplitudes so that as many Hermite function coefficients as possible vanish. The first few relations are

$$a_k - 2^{-1} r_1 = 0, \tag{7.47}$$

$$2^{-1/2} r_1 - 3^{-1/2} r_3 = 0, \tag{7.48}$$

$$2^{-1} r_3 - 5^{-1/2} r_5 = 0, \tag{7.49}$$

and so forth. Only amplitudes of Rossby waves with odd indices figure into this condition because these are the ones that are symmetric about the equator, and therefore interact with the Kelvin wave. Amplitudes of antisymmetric waves vanish at $x = x_{\mathrm{E}}$. Since only a finite number of waves can be considered, the condition that there be no flow into the boundary cannot be exactly satisfied; there will be a component of normal flow into the boundary stemming from the highest meridional mode Rossby wave in the simulation.

At the western boundary the no normal flow condition cannot be met by a long-wave model, since satisfaction of the boundary condition requires the short waves, which are not present. The proper course of action (see, e.g., Cane and Sarachik, 1977) is to impose the condition that there be no net flow into the western boundary, i.e. $\int_{-\infty}^{\infty} u(x_W, y)\mathrm{d}y = 0$. Symmetric (i.e., odd order) Rossby waves impinging on the western boundary reflect as Kelvin waves, with amplitudes that can be derived by integrating the u component of (7.44) with respect to y from $-\infty$ to ∞, with the result that

$$\int_{-\infty}^{\infty} u_{\text{Kelvin}}\,\mathrm{d}y = \pi^{1/4} a_k, \tag{7.50}$$

and

$$\int_{-\infty}^{\infty} u_{\text{Rossby}}\,\mathrm{d}y = -\pi^{1/4}\alpha_n[2n(n+1)]^{-1} r_n, \tag{7.51}$$

for the nth Rossby mode, where $\alpha_n = [2^n n!]^{-1}[(2n+1)!]^{1/2}$. We are now in a position to describe a simple model of a tropical ocean, whose motion is governed by the linear shallow-water equations on the equatorial β-plane, infinite in meridional extent, bounded at the east and west by meridional walls.

Beginning with initial conditions for all wave amplitudes in the domain, integrate (7.45) along its characteristics from west to east. This results in a new set of Kelvin wave amplitudes at the advanced time. The Kelvin wave amplitude at the eastern boundary can then serve as a boundary condition for (7.46), which can then be integrated westward from $x = x_{\mathrm{E}}$ to $x = x_{\mathrm{W}}$ to complete the step. The Kelvin wave amplitude at $x = x_{\mathrm{W}}$ is determined by (7.50)–(7.51).

V can be retrieved diagnostically from U and P by eliminating the time derivatives in (7.39)–(7.41) to form

$$V_{yy} - y^2 V = -\tau^{(y)} + y\tau^{(x)} - yP_x - U_{xy}. \tag{7.52}$$

Simple wave models of the tropical ocean are attractive because they can represent much of the variability of the tropical oceans on large spatial and temporal scales in economical form. Kawabe (1994) was able to represent much of the variability of the tide gauge data in the tropical Pacific in terms of equatorial Rossby and Kelvin waves. The economy afforded by simple wave models allows experimentation with advanced data assimilation methods that might be prohibitively expensive in terms of computational resources; see, e.g., Miller and Cane (1989, 1997), Chan *et al.* (1996).

Extensions of the theory to more complex boundary geometry have been formulated: du Penhoat *et al.* (1983) extended the theory to partial boundaries, so simple models of the tropical Atlantic, with schematic representations of Brazil and the Gulf of Guinea could be cast in this form. They also showed that the presence of islands in the tropical oceans made little difference in the propagation of long Kelvin and Rossby waves. These extensions of the simple wave theory come at significant cost in complication.

The usefulness of the simple wave formulation as an explicit model of the real ocean is limited by a number of factors. One consideration is the slow convergence of Hermite series. For this reason, simple wave models lose precision rapidly away from the equator. With scales typical of the tropical Pacific, the most poleward root of the Hermite function of degree six lies about 6° from the equator. This means that a simple wave model that includes five Rossby modes will be valid roughly between 6° S and 6° N. Larger domains and better resolution can be had with the inclusion of more Rossby modes, but troubles arise as the number of meridional modes increases. The forcing at each longitude must be projected on the Hermite modes, and this involves a quadrature over the entire meridional range of the domain. Higher Rossby modes require more highly resolved forcing data along the length of meridians, and this may not be available. From a longer point of view, it seems clear that outside of a small band of latitudes near the equator, response of the ocean to wind forcing becomes more local in nature, and is not efficiently represented in terms of long waves.

7.3.1 A simple gridded model of the tropical ocean

In this section we present an example of a simple gridded shallow-water model, that of Cane and Patton (1984; hereafter CP). Gridded models do not suffer from the convergence problems affecting simple wave models, so they will be more efficient than simple wave models in domains greater than a few degrees in meridional extent. This model is implemented in the form of a numerical solution to (7.39)–(7.41), the shallow-water equations for each baroclinic mode, subject to the long-wave approximation. The properties of the wavelike solutions to (7.39)–(7.41) are considered explicitly in the implementation of the model. In this model, the velocity components and layer thickness appear on a grid, there are no long Hermite series that must be added up explicitly, and one is spared most of the quadratures along meridians. The details of the design of this model are presented here, since they contain much that is illustrative of the basic principles of numerical modeling. This model has been widely applied in studies of the tropical ocean, and it forms the oceanic component of the Zebiak–Cane coupled model of the tropical ocean and atmosphere that has been used to predict the ENSO cycle; see Zebiak and Cane (1987).

This is a gridded model, so boundaries may be reasonably faithfully represented, but since there are no natural bounds on the meridional extent of the domain, the model allows for the placement of artificial zonally oriented barriers to form the poleward boundaries. There is obvious inefficiency in defining a model domain that extends from the Antarctic to the Bering Sea when one's major interest is in the tropical Pacific, even though a case could be made for modeling regions poleward of, say, 30° with very coarse resolution.

The imposition of artificial barriers leads to physical consequences. A model with solid boundaries to the north and south admits westward propagating solutions of the form (7.11). These are sometimes called the "anti-Kelvin waves," since they share many properties with the equatorial Kelvin waves. These waves do not occur in nature, but will exist in a model of limited meridional extent.

The equatorial Kelvin wave is the only eastward-propagating wave admitted by the system of equations (7.39)–(7.41). In the CP model, the solution to the model equations is separated into two parts, one consisting of the Kelvin wave and the other consisting of the long Rossby waves and the anti-Kelvin wave. The amplitude of the Kelvin wave is calculated separately by integrating (7.45) eastward along its characteristics. The boundary condition is determined by mass balance at the western boundary according to (7.50)–(7.51). At partial boundaries, transmission coefficients for the waves are calculated according to formulas derived by du Penhoat *et al.* (1983).

Boundary conditions at the eastern boundary are determined by the Kelvin wave amplitude. Beginning at the eastern boundary, the meridional velocity component v is calculated according to (7.52), and equations similar to (7.39)–(7.41) are solved westward and forward in time (i.e., upwind with respect to westward-propagating waves). The difference between the equations solved in this phase of the calculation and (7.39)–(7.41) is the explicit subtraction of Kelvin wave forcing from the right-hand side, and the inclusion of weak Rayleigh friction.

This scheme is downwind, and therefore unstable with respect to the Kelvin waves, and while the Kelvin wave forcing has been removed explicitly, Kelvin waves will eventually be excited by noise. For this reason, u and h fields with the exact finite-difference form of the Kelvin wave are explicitly filtered at each time step, at the same time the Kelvin wave forcing is filtered.

7.3.2 Example: Application of a simple model of the tropical Pacific Ocean

One common conceptual model of the El Niño–Southern Oscillation (ENSO) cycle is the *delayed oscillator*. In this model, a limited region of anomalous westerly winds arises in the central Pacific through some ocean-atmosphere instability. These winds cause the free surface to slope upward and the thermocline to slope downward to the east, while the free surface slopes downward and the thermocline slopes upward to the west. The disturbance propagates eastward as a downwelling Kelvin wave, and westward as an upwelling Rossby wave; a deepening (shoaling) thermocline is the way the warming (cooling) is represented in models of this kind. In the simplest and purest form of the delayed oscillator theory, the reflection of the Rossby wave from the western boundary gives rise to an upwelling Kelvin wave, which then propagates to the region of anomalous westerlies and counteracts the instability. There are many obvious questions about this theory, which has spawned an extensive literature; see, e.g., Neelin *et al.* (1998) and references therein. A glance at a map will reveal the obvious question of whether a simple meridionally oriented wall is a reasonable boundary condition. The role of off-equatorial motions is also in question.

Wakata and Sarachik (1991) used the CP model to investigate the question of whether a model ocean response to observed wind was consistent with the delayed oscillator hypothesis. They used the FSU wind data set (Stricherz *et al.*, 1992) to force the model for 28 years, and diagnosed the model output in terms of Kelvin and Rossby modes.

Figure 7.2 shows the evolution of the model thermocline depth on the equator from 1961 to 1988. The broad shape of the equatorial Pacific thermal structure is clearly visible, with the thermocline lying deep west of the dateline and shoaling gently to the east. East of the dateline, the thermocline shoals more steeply. The ENSO events are clearly visible with the downwelling Kelvin waves propagating rapidly eastward to signal the onset of the warming event in the eastern part of the basin. Since the Kelvin waves cross the basin in just over two months, the lines on

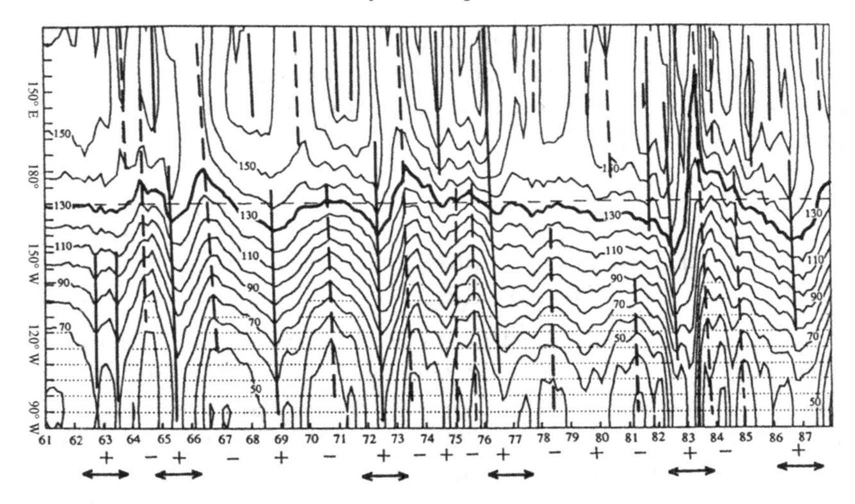

Fig. 7.2 Thermal response of the shallow-water model at the equator. Contour map in longitude vs. time (years) of thermocline depth. Regions where the thermocline is shallower than 70 m are dotted. Heavy solid and dashed lines show crests of downwelling and upwelling Kelvin modes respectively. Reproduced from Figure 1 of Wakata and Sarachik (1991), with permission of the American Meteorological Society. Arrows below the abscissa mark warm events. "+" and "−" signs mark transition of the 130 m isobath across 168° W.

this graph that trace the propagation of the waves are nearly vertical and the propagation is barely discernible.

The notable feature of the propagation of the Kelvin wave crests is that most of the ones that originate on the western boundary, presumably from reflection of Rossby waves of like sign, propagate to the central Pacific and no farther, usually not much past the dateline. With few exceptions, the Kelvin modes that correspond to the warm and cold events in the eastern Pacific originate in the central part of the basin, most just east of the dateline.

Wakata and Sarachik studied the Kelvin and Rossby waves in terms of our variable $q = U + P$ (see (7.23)), which evolves according to the inhomogeneous form of (7.25):

$$q_t + q_x + V_y - yV = \tau^{(x)}. \tag{7.53}$$

The Kelvin and Rossby wave-forcing functions may be expressed in terms of the coefficients $\hat{\tau}_n(x, t)$ in (7.43). If we expand q in Hermite functions,

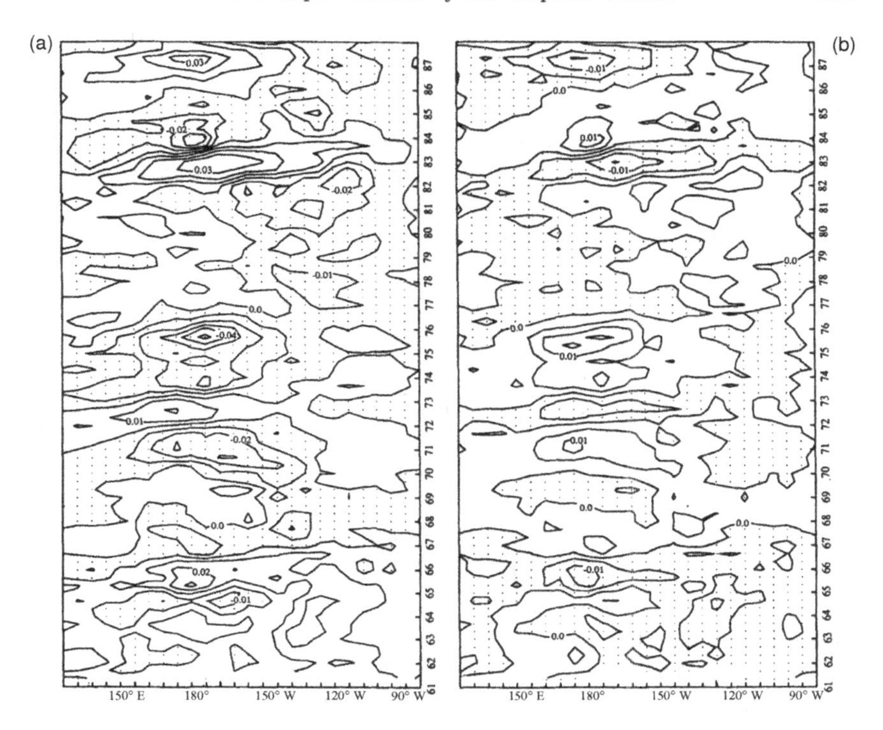

Fig. 7.3 Wind forcing projected onto meridional modes. (a) Anomaly relative to the seasonal cycle of the Kelvin mode forcing. Negative area is dotted. (b) Anomaly of the first symmetric equatorial Rossby mode forcing. Reproduced from Figure 2 of Wakata and Sarachik (1991), with permission of the American Meteorological Society.

$$q = \sum q_n(x,t)\psi_n(y), \tag{7.54}$$

we find that the Kelvin wave component q_0 evolves according to

$$q_{0t} + q_{0x} = \hat{\tau}_0, \tag{7.55}$$

where $\hat{\tau}$ is the coefficient of Ψ_0 in the Hermite expansion of $\tau^{(x)}$ and the Rossby wave components q_n evolve according to

$$q_{nt} - \frac{1}{(2n+1)} q_{nx} = \frac{n}{(2n+1)} \hat{\tau}_{n+1} - \frac{(n^2+n)^{1/2}}{(2n+1)} \hat{\tau}_{n-1} \tag{7.56}$$

(see Exercise 7.1).

Kelvin and Rossby wave-forcing functions, calculated by application of numerical quadratures to the FSU wind data set, are shown in Figure 7.3.

Positive Kelvin wave forcing leads to an eastward-propagating disturbance that elevates the surface and depresses the thermocline. This is a *downwelling* Kelvin wave, and it is associated with warming of the upper ocean. Strong downwelling Kelvin waves associated with westerly wind anomalies along the equator, i.e., weakening of the trade winds, are clearly evident near the times associated with ENSO events of 1966–7, 1972–3, 1982–3 and 1986–7. These regions of positive Kelvin wave forcing are concentrated near the dateline and extending eastward, as the theory would require, and forcing is weak east of 150° E.

Forcing of the first mode symmetric Rossby wave shows similar structure but opposite sign from the Kelvin wave forcing. Referring to (7.56), and noting that $\hat{\tau}_2 << \hat{\tau}_0$, we see that the forcing of this wave is approximately proportional to the Kelvin wave forcing with opposite sign; from (7.56), the constant of proportionality is $(-2^{1/2}/3)$. Physically, this corresponds to the fact that an isolated patch of westerly winds will force downwelling waves eastward and upwelling waves westward, in order that the themocline will adjust downward to the east and upward to the west; see, e.g., Philander (1991).

Figure 7.4 depicts the model Kelvin wave anomaly amplitude, with paths of Rossby wave crests superimposed. This figure is complementary to Figure 7.2, in that it shows the evolution of the wave amplitudes as opposed to the thermocline depth itself. Here we see another view of the overall scheme of the delayed oscillator. A westerly anomaly, i.e., a weakening of the tradewinds, near the dateline gives rise to a downwelling Kelvin wave that propagates east and an upwelling Rossby wave that propagates west. The reflection of the Rossby wave results in an upwelling Kelvin wave that changes the air–sea interaction near the dateline in such a way that an anomaly of opposite sign is produced, and an upwelling Kelvin wave propagates eastward, resulting in the cold phase of the ENSO cycle, often called "La Niña" (see, e.g., Philander, 1991).

The main point of this work can be summarized in terms of the wave paths shown in Figures 7.2 and 7.4. The motion of the crests of the Kelvin waves is shown as heavy solid and dashed lines in Figure 7.2. From these, we see that the heavy solid lines that originate at the eastern boundary, in this case the top of the figure, do not line up with those that reach the eastern boundary, signaling the onset of an El Niño event. In Figure 7.4, upwelling Rossby waves, forced by the same westerly anomalies that force the downwelling Kelvin waves (see

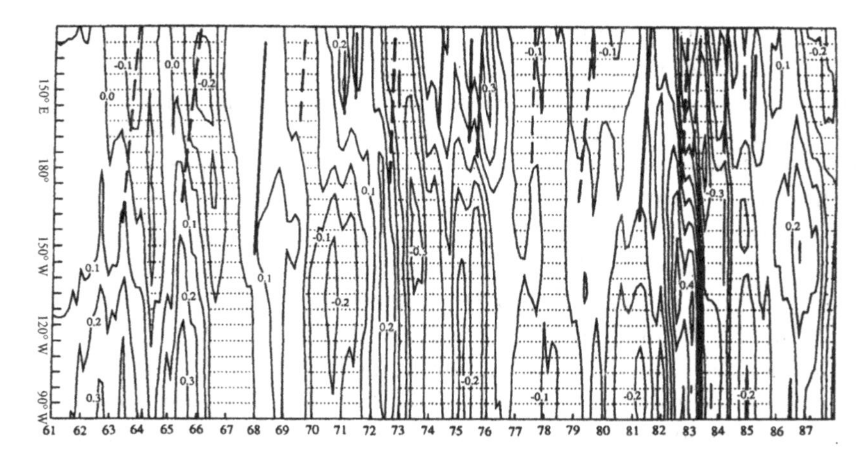

Fig. 7.4 Kelvin wave anomaly amplitude, with peaks of first mode symmetric Rossby wave superimposed. Dotted area shows upwelling. Heavy solid (dashed) lines denote peaks of downwelling (upwelling) first symmetric mode Rossby waves. Reproduced from Figure 3 of Wakata and Sarachik (1991), with permission of the American Meteorological Society.

Figure 7.3), propagate freely westward, there to result in upwelling Kelvin waves.

In summary, while linear shallow-water models like the CP model have difficulty in representing the thermodynamic aspects of tropical ocean circulation, this study showed that the response of a shallow-water model to observed winds is consistent with the delayed oscillator hypothesis.

While the CP model is a gridded model, which represents Kelvin wave dynamics explicitly, but not Rossby wave dynamics, the results were described by Wakata and Sarachik (1991) entirely in terms of wave propagation. Similar results could have been obtained from a simple wave model. One of the most controversial aspects of the delayed oscillator theory is the assumption of perfect reflection of Rossby waves at the western boundary. Models such as this one in which the long-wave approximation is imposed lack the capability of reproducing the details of the flow near the western boundary, since this requires the short Rossby waves with eastward group velocity, and the long-wave approximation filters these waves. Some assumption about the reflection of waves at the western boundary must therefore be made, and the assumption of perfect reflection is therefore a feature of the model rather than a result.

7.3.3 Simple models of the tropical Atlantic Ocean

Early theoretical work described the variability of large-scale low-frequency response of the Atlantic Ocean to wind forcing in terms of long waves (e.g., Cane and Sarachik, 1977, 1981), and evidence of equatorially trapped long waves in the Atlantic Ocean is widely reported (e.g. Weisberg *et al.*, 1979), but there have been few numerical modeling studies of the tropical Atlantic based on explicit wave dynamics, almost certainly due to the fact that the boundaries cannot be reasonably represented in terms of meridional walls. A glance at the map of the Pacific reveals that the boundaries are not well described by meridional walls in that ocean either, but the spatial extent of the Pacific is so much greater than that of the Atlantic that more of the Pacific can be considered to be far from boundaries, and the details of boundary interactions can be considered relatively less important for physical variability in mid-ocean in the Pacific than in the Atlantic.

The simplest models of the equatorial Atlantic Ocean have been linearized shallow water models implemented on grids. Du Penhoat and Treguier (1985) implemented the CP model for the tropical Atlantic. Their implementation included nine baroclinic modes, of which the first three accounted for the bulk of the variability. The horizontal resolution was 0.75° in the zonal direction and 0.45° in the meridional. The radii of deformation for the third mode and fourth mode were 1.80° and 1.51° respectively, so motion at the higher modes would not have been well resolved.

The Brazilian coast and the Gulf of Guinea were represented by rectangular sections cut out of the lower left and upper right corners respectively. Artificial zonal walls are imposed at 18° north and south. The model was driven for four years with the Hellerman–Rosenstein monthly wind data set (Hellerman and Rosenstein, 1983) and results were presented for the fourth model year. The Kelvin wave speed for the third mode was given as $0.91\,\mathrm{m\,s^{-1}}$ and the basin is 60° wide at its widest point, so three years should be long enough for the system to be reasonably well spun up. The ninth mode, with a Kelvin wave speed of $0.27\,\mathrm{m\,s^{-1}}$ probably wouldn't have had time to spin up, but the higher modes were fairly heavily damped, so much so that the damping time would be less than the time required for a disturbance at the western boundary propagating eastward as a Kelvin wave, and the effect of that disturbance returning from the eastern boundary with the fastest reflected Rossby wave.

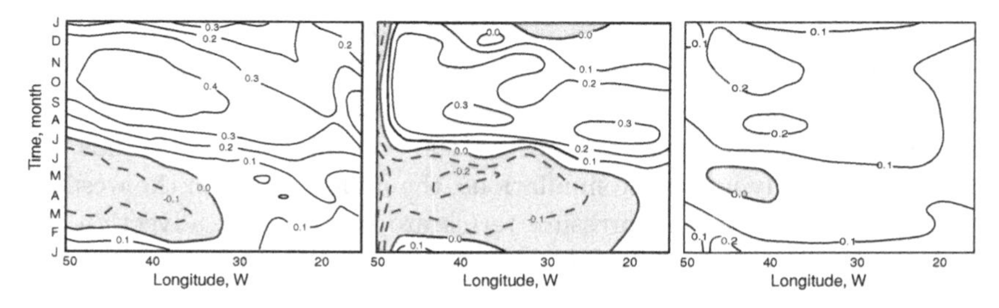

Fig. 7.5 Annual cycle of zonal currents in a latitude band just north of the equator. Left to right: model output, averaged from 5° N to 8° N; observed ship drift data (Richardson and McKee, 1984), averaged from 5° N to 8° N; geostrophic velocity derived from 500 dbar dynamic height data, calculated from hydrographic data. Left-hand and center panels redrawn from Figure 7 of du Penhoat and Treguier (1985), by permission of Elsevier; right-hand panel redrawn from Figure 4 of Arnault (1997), by permission of the American Geophysical Union. Contour interval = $0.1\,\mathrm{m\,s^{-1}}$. Shaded areas represent negative (westward) velocities.

The model reproduced the ridges and troughs that characterize the dynamic topography of the tropical Atlantic remarkably well for the most part, the exception being the region south af about 5° S and west of about 10° W, where wind data are particularly sparse. The model does fairly well in the southeastern part of the basin, despite problematic winds, probably due to the influence of reflected Rossby waves from the boundary. The basic features of the phase of the annual cycle are also present, as can be seen in the left-hand and center panels of Figure 7.5.

The North Equatorial Countercurrent (NECC) compares remarkably well to ship drift observations east of about 48° W as reported by Richardson and McKee (1984) and shown here in the center panel of Figure 7.5. West of that point, the flow is dominated by boundary effects which are not treated by the model by virtue of the long-wave approximation. It is curious that Du Penhoat and Treguier note that the model NECC is geostrophically balanced for the most part, but the model actually shows better agreement with the ship drift data, shown in the center panel of Figure 7.5, than it does with the geostrophic calculation of Arnault (1997), shown in the right-hand panel.

This linear model in fact produces an equatorial undercurrent. In the model undercurrent, the first three modes, which dominate other aspects of the seasonal variability, contribute only 25% of the mean transport. The model undercurrent suggests the presence of such a current, but does not compare well with observations. Its seasonal variability is out of phase with the observed, and its intensity decreases too rapidly to the east. In this linear context, the undercurrent will be governed by a balance between the negative pressure gradient, i.e., the upward slope of the surface to the west under the influence of the prevailing easterly winds, and friction. In fully nonlinear simulations, the undercurrent in the west is strongly accelerated, and pressure terms are balanced by advection, which has greater influence than friction. A model undercurrent that is more faithful to observations requires explicit inclusion of advection.

7.4 Application of general circulation models

7.4.1 Comparisons to simple models

Boulanger *et al.* (1997) extracted the first baroclinic mode Kelvin and Rossby wave amplitudes from a general circulation model (GCM) simulation and compared them to the results of calculations performed with an explicit long-wave model. The GCM was a C-grid model with 30 levels in the vertical direction. The vertical grid spacing was set to 10 m for the upper 10 levels, increasing gradually below the tenth level. Meridional resolution was 0.33° at the equator, increasing gradually to 1.5° at the artificial northern and southern boundaries imposed at 50° N and 50° S. The zonal resolution near coastlines was 0.33°, increasing to 0.75° in the interior of the basins. The model domain included all three tropical oceans. Time stepping for the dynamical equations was mostly by leapfrog, with the usual exceptions for the diffusion terms. Turbulent horizontal diffusion terms were treated explicitly, while turbulent vertical terms were treated implicitly.

Because this model extends to the bottom, it will experience some climate drift unless there is some provision for simulating the thermohaline circulation. This is done by relaxing the model state to the Levitus (1982) climatology poleward of 20° latitude. Wind forcing and heat flux were provided by the ARPEGE climate model (Déqué *et al.*, 1994). The ARPEGE heat flux was modified by the addition of relaxation to observed SST (Reynolds and Smith, 1994), so the heat flux Q was governed by

$$Q = Q_{\text{Arpege}} + K_Q(SST - SST_{\text{obs}}), \tag{7.57}$$

with $K_Q = -40\,\text{W}\,\text{m}^{-2}\,\text{K}^{-1}$.

In order to project the output of the GCM upon the first baroclinic mode Kelvin and Rossby waves, it is necessary to estimate a speed. This was done by time lag autocorrelation analysis at the equator and at 4° N. The result was an estimated speed of $2.5\,\mathrm{m\,s^{-1}}$, which in turn leads to a deformation radius of 332 km. Given an explicit estimate of the deformation radius, Rossby and Kelvin wave amplitudes can then be calculated explicitly by quadratures. The Kelvin wave speed was estimated to be $2.20 \pm 0.5\,\mathrm{m\,s^{-1}}$ and the lowest meridional model Rossby wave speed was estimated to be $-0.80 \pm 0.5\,\mathrm{m\,s^{-1}}$, consistent with the earlier estimate and with the theoretical prediction that the lowest mode Rossby wave speed would be $-1/3 \times$ the Kelvin wave speed.

Summary statistics for the ten-year run show that much of the wind forcing in the equatorial waveguide can be accounted for by forcing of the Kelvin and first symmetric Rossby modes. There are small regions near 150°E and 90°W where forcing of these two modes only accounts for about half the total wind variance within a deformation radius of the equator, but over most of the basin, these two modes account for about 60–80% of the total wind forcing variance about the seasonal cycle. About 90% of the dynamic height anomaly within the waveguide can be explained by these two modes.

The GCM results were then compared to a very simple wave model based on equations differing from (7.45)–(7.46) only by a linear damping term of 3 months^{-1} for the Kelvin and first Rossby modes. There were no reflections in the model, only wind forcing. The result from the simple model matched the Kelvin wave amplitude extracted from the GCM reasonably well from just east of the dateline to about 110° W. Rossby wave amplitudes match rather less well. The correlation reaches a peak of about 0.9 west of the dateline but the amplitudes do not match very well.

In the last decades data with extensive coverage of dynamic height in the Pacific from the TAO array and from satellite altimetry have become available, and it is now possible to extract wave properties from data alone. The speeds calculated from data are rather slower than would be predicted by the shallow-water theory covered in Section 7.2, though not inconsistent with earlier estimates due to the large error bars associated with those earlier estimates due to the relative scarcity of data. The observed waves also lack the symmetry properties of the Hermite functions, probably due to background baroclinic currents. The reader should not be surprised that a theory based on small disturbances about a uniform state of rest is not correct in detail. A thorough review of this topic can be found in Chelton *et al.* (2003).

Perez *et al.* (2005), in their modeling study, found asymmetry of Rossby wave amplitudes similar to the observed, viz. with greater amplitude north of the equator. In processing the model output similar to the way the altimeter data were processed for Chelton *et al.* (2003), Perez *et al.* were able to extract wave speeds similar to those observed. This is evidently due to the background flow. Experiments with artificially symmetrized background flow did not yield the observed result, even with realistic asymmetrical forcing.

Purely data-based calculations of the propagation speed of the annual Rossby wave in the Pacific show the curious result that the waves are faster north of the equator than south. When Perez *et al.* processed their model output in a fashion similar to the way the data had been processed they also observed this phenomenon, but it disappeared in a similar calculation in which the boundary was forced to be exactly rectangular; evidently the asymmetry in speed, if not in amplitude, stems from the asymmetry of the domain.

7.4.2 Example: Dynamic balances in model equatorial undercurrents

The equatorial undercurrent is one of the most prominent features of equatorial ocean circulation. Over much of the tropical oceans, the trade winds set up a pressure gradient that tends to drive a strong eastward current, in opposition to the westward current at the surface. This equatorial undercurrent can have speeds of the order of $1\,\mathrm{m\,s^{-1}}$ at its core, and transports in the range of tens of Sverdrups ($1\,\mathrm{Sv} = 10^6\,\mathrm{m^3\,s^{-1}}$). Over the years a variety of dynamical explanations have been proposed for the existence of the undercurrent. Wacogne (1989) and Wacogne (1990) contain a description of a series of experiments in which a GCM was driven with seasonal winds, and individual terms in the fully nonlinear momentum balance equations were evaluated in order to assess the importance of different physical effects.

The model was an adaptation of the Bryan–Cox (cf. Cox (1984)) model to the tropical ocean (Philander and Pacanowski, 1984, 1986a,b; Philander *et al.*, 1986). The model was driven by the Hellerman and Rosenstein (1983) wind and heat flux data set. Surface stress is imposed, and air temperature is specified at every step, but no surface boundary condition is imposed on salt flux at the surface to take

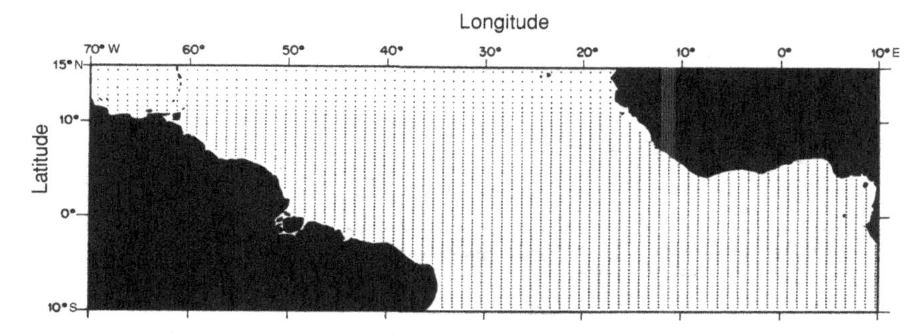

Fig. 7.6 Grid and basin geometry for the model of the equatorial Atlantic, in the model described by Wacogne (1989). Near the artificial zonal walls at 30° N and 50° S artificial buffer zones are introduced in which the solution is forced toward the local climatology.

account of evaporation and precipitation. Accurate simulation of salinity is therefore not expected near the surface. The model was configured with 27 levels in the vertical, with 10 m resolution above 100 m. The horizontal resolution in the equatorial region was 1° longitude and 1/3° latitude. Vertical diffusion was parameterized in terms of Richardson number according to the recipe of Pacanowski and Philander (1981). A constant horizontal eddy viscosity of $2 \times 10^3\,\mathrm{m^2\,s^{-1}}$ was imposed. The model domain is shown in Figure 7.6.

The seasonal cycle spins up in only about two years. Mean annual results are summarized in Figures 7.7 and 7.8. The mean wind stress is easterly west of 8° W and westerly east of 8° W. The meridional wind is southerly throughout the basin.

The plot of zonal velocity versus longitude shows that the undercurrent shoals eastward as expected, with the core above the thermocline in the west and below the thermocline in the east. Above the core there is a layer of high-shear fluid about 50 m thick, above which the water flows westward. Referring to the plots of zonal and vertical velocity versus longitude (second and third panels from the top in Figure 7.7), if we estimate the zonal velocity at $0.5\,\mathrm{m\,s^{-1}}$ and the vertical velocity at $2 \times 10^{-5}\,\mathrm{m\,s^{-1}}$ we find that it takes 2.5×10^6 s, about a month, for a water parcel to rise 50 m through the high shear layer to the westward-flowing surface water, during which time the fluid travels $1.25 \times 10^6\,\mathrm{m} \approx 10°$, about 1/5 of the width of the basin. It therefore appears that no fluid parcel traverses the entire basin intact within the undercurrent, and water must be fed into the undercurrent by the meridional flow.

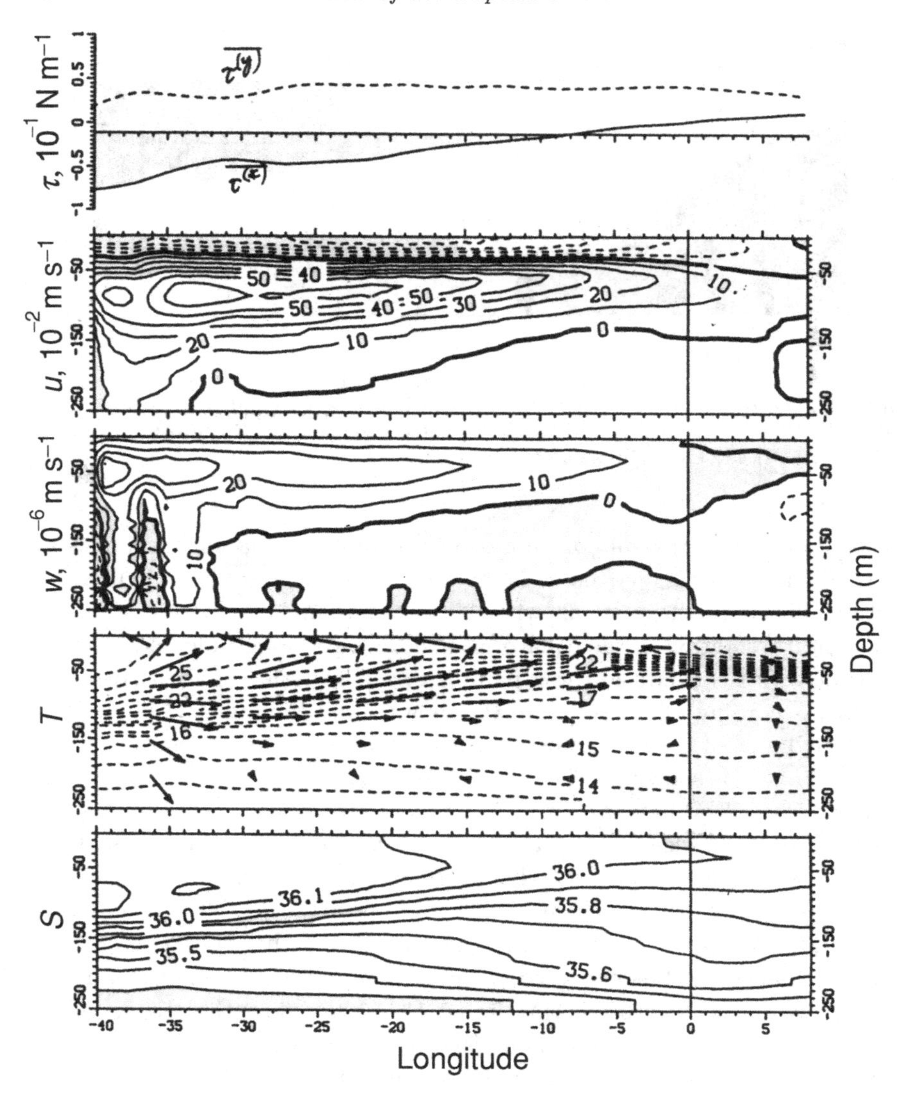

Fig. 7.7 Longitude–depth sections of annually averaged quantities. Top to bottom: zonal (solid) and meridional (dashed) wind stress components; zonal and vertical velocity components, resp.; temperature. Redrawn from Figure 2 of Wacogne (1989), with permission of the American Geophysical Union.

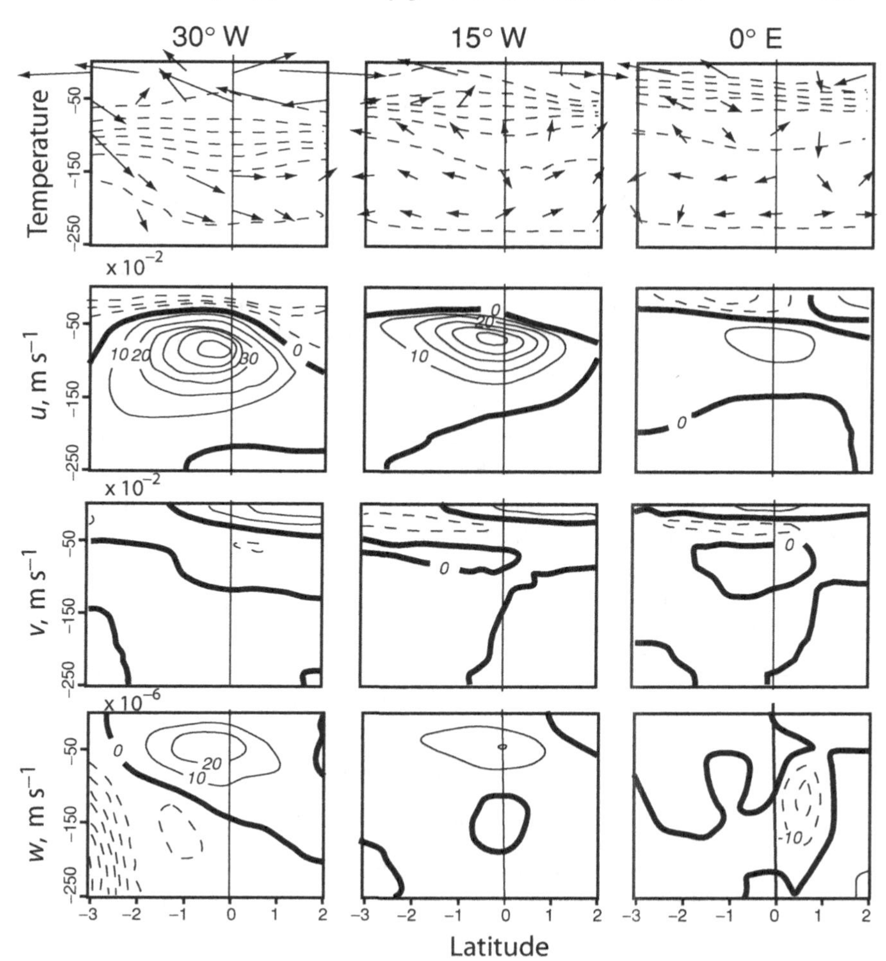

Fig. 7.8 Latitude–depth sections of annually averaged quantities. Top to bottom: temperature, zonal velocity, meridional velocity, vertical velocity. Redrawn from Figure 3 of Wacogne (1989), with permission of the American Geophysical Union.

By 30° W the undercurrent is fully developed, and starting to decelerate. The acceleration region is narrow and confined west of 33° W. Shoaling and deceleration of the undercurrent core is evident in Figure 7.8. Comparison of the temperature section in the upper panel with the zonal velocity section in the panel below it shows the weakening of the undercurrent as it progresses eastward. This weakening is evident both from the speeds themselves and the slopes of the isotherms. Note also, from these sections as from Figure 7.7, that the undercurrent core is above the thermocline in the west and below the thermocline in the east.

The plot of the meridional velocity shows the meridional flow to be divergent at the surface. As we expect poleward flow, i.e., divergence, near the surface we expect the geostrophically balanced equatorward flow just off the equator to be convergent at depth. This expected convergence at depth may be present, but is not so readily apparent in this plot. Taken together, the plots of temperature and the three velocity components form a picture of displacement of the surface divergence south of the equator, and corresponding northward displacement of convergence at depth. This may be a consequence of the southerly component of the mean wind; see the upper panel of Figure 7.7.

We are left with the overall picture that water comes into the undercurrent from the west and north at all longitudes, and from the south at longitudes west of 15° W. Water leaves the undercurrent by upwelling into the westward flow, and, to a lesser extent, downwelling into an underlying deep westward flow, veering southeastward east of 0° E.

In the mean zonal momentum balance, the zonal advection uu_x was only significant west of 35° W. For most of the upper thermocline (i.e., above 75 m or so) and east of 30° W along the equator, the mean annual zonal momentum balance was given approximately by

$$\overline{vu_y} + \overline{wu_z} \approx \overline{-p_x/\rho_0} + \overline{(\nu u_z)_z} < 0, \tag{7.58}$$

so the pressure gradient cannot overcome friction and the undercurrent decelerates eastward, as we have seen. Off the equator, between 1° and 2°, coriolis becomes most important.

As this paper was written, there was little data available in the tropical Atlantic, but the model output exhibited a number of the observed features, e.g., the deceleration of the undercurrent, and the appearance of the undercurrent core above the thermocline in the west and below the thermocline in the east.

The results here support a view of the undercurrent in which the different theories are not so much competitors, as ways of understanding the undercurrent in different regimes, e.g., Pedlosky's (1987) inertial theory may apply to the inertially accelerated regime west of 35° W.

In what might be considered a companion study to Wacogne (1989), Maes *et al.* (1997) performed a series of experiments with a GCM developed at the Laboratoire d'Océanographie Dynamique et de Climatologie

(LODYC) in which they simulated the wind-forced behavior of the tropical Pacific Ocean on a grid with horizontal resolution at the equator of 0.75° and 0.5° in the zonal and meridional directions respectively, and 20 levels in the vertical. The meridional grid spacing increases to 2° at the artificial northern and southern boundaries, imposed at 30° N and 30° S. Surface forcing was given by the Hellerman and Rosenstein (1983) data set. Surface heat flux was taken from Oberhuber's (1988) atlas, with additional relaxation to Oberhuber's SST analysis. Surface salinity was relaxed to Levitus (1982) surface salinity climatology. Vertical diffusion was parameterized according to a turbulence closure model. They performed three simulations, differing only in their choices of horizontal diffusivity in the 10° S–10° N region. Their high "H," medium "M" and low "L" eddy viscosities were chosen as $10^4\,\mathrm{m^2\,s^{-1}}$, $10^3\,\mathrm{m^2\,s^{-1}}$ and $10^2\,\mathrm{m^2\,s^{-1}}$ respectively.

The found their model undercurrent to be quite sensitive to the value of the horizontal diffusivity. In their intermediate and low diffusivity runs, the model undercurrent exhibited a dynamic balance in the upper thermocline similar to (7.58), the major difference being that zonal advection was more important in the Pacific simulation, while meridional advection was negligible. In the high diffusivity case H, the balance was among zonal pressure gradient, vertical advection and horizontal and vertical diffusivity. Horizontal advection was relatively unimportant, and the model undercurrent was much weaker than it was in the M and L runs.

The question of how to choose a value for the horizontal diffusivity is not a simple one, since it has implications for the overall dynamic balances. There are clearly problems with picking the diffusivity too large, as in the H run of Maes *et al.* (1997), but picking a small value of the horizontal diffusivity tends to increase the effect of vertical diffusivity. Fortunately there is not that much difference between the M and L runs of Maes *et al.* (1997), so the system is not very sensitive to variations in that range, but a better understanding of the small-scale processes that transport mass and momentum is certainly needed.

7.5 Conclusions

The success of early modeling efforts could be attributed in large part to the nature of the problem: motion on the scales of interest could be described, to the level supported by available data, by simple dynamics. Models based on linearized dynamics designed to describe anomalies about a poorly known mean proved to be useful in modeling interannual variations in the Pacific and seasonal variations in the Atlantic. The Indian Ocean was more complex, and much of the interesting variability was out of the waveguide, but comparatively simple models, e.g., single-layer reduced-gravity models, could explain most of the available observations.

New extensive data sets have become available, beginning at the end of the last century and continuing to the present. With much better forcing fields and data available from in-situ and remote-sensing platforms, the demands upon the current and succeeding generations of models of the tropical ocean will be much more exacting.

One might think that solution of remaining problems with simulation of the tropical ocean await the computing resources to refine the grids significantly beyond the 1° zonal, 1/3° meridional and 10–20 vertical layers that represent the nominal resolution of most GCMs as this is being written. More resolution might well bring advances in understanding, but the nominal resolutions in use today do a good job of resolving motion at the scale of the equatorial deformation radius and vertical grids with 20 levels or so in the upper ocean are not uncommon. This should be sufficient resolution to simulate the scales of interest, yet the systematic difficulties remain.

Simulations initialized with atlases of climatology such as Levitus (1982) and driven with available wind products will tend to drift away from the climatology. This is to be expected; the climatology and the forcing data sets have errors of their own, and even a perfect model would not produce the climatology from Levitus (1982) in response to forcing by any available collection of wind and heat flux products.

Comparisons among the results of driving models with different wind products have been performed by Hackert *et al.* (2001) and Perez (2005). No single wind product emerged as the clear best choice for all simulations.

Systematic errors remain in GCM simulations of the tropical ocean for all wind products. At this point factors other than wind-stress errors may make significant contributions to the errors in the model outputs. It is likely that, in the tropics, horizontal grid spacing of 0.25° to 0.33° in models with 20–30 layers or levels in the vertical direction provide sufficient resolution for most features of the tropical ocean, and we must look elsewhere for sources of systematic error. Likely candidates are errors in buoyancy fluxes and in parameterization of small-scale mixing processes.

We now have enough observations in long enough time series to warrant the use of detailed models. The detailed models certainly have the potential to produce better analyses than the simple models, but two decades ago sufficient data to make the distinction did not exist. The challenge now is to implement models that will produce accurate estimates of the mean and seasonal structures of the tropical oceans. This necessarily involves estimating the biases and understanding their physical origins. In the highly nonlinear setting of GCMs, these are the major obstacles to applying GCMs to advancement of understanding of the tropical oceans and to operational uses such as climate prediction.

7.6 Exercises

7.1 *Forcing of Rossby waves*

(a) Use the methods of Section 7.2 to derive (7.56). You may proceed as follows:

Begin with the inhomogeneous forms of (7.25) and (7.26):

$$q_t + q_x + V_y - yV = \tau^{(x)}, \tag{E7.1}$$

$$r_t - r_x + V_y + yV = -\tau^{(x)}. \tag{E7.2}$$

Expand V and τ in Hermite functions: $V = \sum V_n(x,t)\psi_n(y)$, $\tau = \sum \hat{\tau}_n(x,t)\psi_n(y)$ and examine a single component V_n. Define $q = Q(x,t)\psi_{n+1}$ and $r = R(x,t)\psi_{n-1}$ and apply (7.31) and (7.32) to find

$$Q_{nt} + Q_{nx} - 2^{1/2}(n+1)^{1/2}V_n = \hat{\tau}_{n+1}, \tag{E7.3}$$

$$R_{nt} - R_{nx} + 2^{1/2}n^{1/2}V_n = -\hat{\tau}_{n-1}. \tag{E7.4}$$

Recall the dispersion relation for Rossby waves in the long-wave approximation: write $Q_n = V_n$, $R_n = CV_n$, and choose C so that the advection speed is correct, i.e.,

$$\frac{n^{1/2} - C(n+1)^{1/2}}{n^{1/2} + C(n+1)^{1/2}} = \frac{-1}{2n+1}. \tag{E7.5}$$

The desired result then follows from (E7.3) and (E7.4).

(b) Find a wind data set and calculate the Rossby wave forcing in the tropical Pacific. Compare your results to Figure 7.3.

7.2 *A simple model*: Implement a simple wave model of a tropical ocean. The Pacific is probably the easiest. Begin with a rectangular basin. Find a wind data set and a sea surface height or dynamic height data set and compare your model results to the data.

References

Allen, J. S. (1980). Models of wind-driven currents on the continental shelf. *Ann. Rev. Fluid Mech.*, **12**, 389–433.

Allen, J. S., Newberger, P. A. and Federiuk, J. (1995). Upwelling circulation on the Oregon continental shelf. Part I: Response to idealized forcing. *J. Phys. Oceanogr.*, **25**, 1843–66.

Allen, M. B. III, Herrera, I. and Pinder, G. F. (1988). *Numerical Modeling in Science and Engineering.* New York: Wiley Interscience.

Anderson, D. L. T., Bryan, K., Gill, A. E. and Pacanowski, R. C. (1979). Transient response of the north Atlantic: some model studies. *J. Geophys. Res.*, **84**, 4795–815.

Arakawa, A. and Lamb, V. R. (1977). Computational design of the basic dynamical processes of the UCLA general circulation model. In *Methods in Computational Physics.* Vol. 17, ed. J. Chang. New York: Academic Press, pp. 174–265.

Arakawa, A. and Suarez, M. (1983). Vertical differencing of the primitive equations in sigma coordinates. *Mon. Wea. Rev.*, **111**, 34–45.

Arnault, S. (1987). Tropical Atlantic geostrophic currents and ship drifts. *J. Geophys. Res.*, **92**, 5076–88.

Atlas, R. M., Hoffman, R. N., Bloom, S. C., Jusem, J. C. and Ardizzone, J. (1996). A multi-year global surface wind velocity dataset using SSM/I wind observations. *Bull. Amer. Meteorol. Soc.*, **77**, 869–82.

Barnier, B., Siefridt, L. and Marchesiello, P. (1995). Thermal forcing for a global ocean circulation model using a 3-year climatology of ECMWF analyses. *J. Mar. Syst*, **6**, 363–80.

Batchelor, G. K. (1967). *Introduction to Fluid Dynamics.* Cambridge: Cambridge University Press.

Battisti, D. S. and Hickey, B. M. (1984). Application of remote wind-forced coastal trapped wave theory to the Oregon and Washington coasts. *J. Phys. Oceanogr.*, **14**, 887–903.

Bennett, A. F. (1976). Open-boundary conditions for dispersive waves. *J. Atmos. Sci.*, **33**, 176–82.

Beckmann, A. and Haidvogel, D. B. (1993). Numerical simulation of flow around a tall isolated seamount. Part I: problem formulation and model accuracy. *J. Phys. Oceanogr.*, **23**, 1736–53.

Ben Jelloul, M. and Huck, T. (2003). Basin-mode interactions and selection by the mean flow in a reduced-gravity quasigeostrophic model. *J. Phys. Oceanogr.*, **33**, 2320–32.

Blayo, E. (2000). Compact finite difference schemes for ocean models. *J. Comput. Phys.*, **164**, 241–57.

Blayo, E. and Debreu, L. (2005). Revisiting open boundary conditions from the point of view of characteristic variables. *Ocean Modelling*, **9**, 231–52.

Bleck, R. (2002). An ocean general circulation model framed in hybrid isopycnic-Cartesian coordinates. *Ocean Modelling*, **37**, 55–88.

Bleck, R. and Smith, L. (1990). A wind-driven isopycnic coordinate model of the north and equatorial Atlantic ocean. 1. Model development and supporting experiments. *J. Geophys. Res.*, **95**, 3273–85.

Blumberg, A. F. and Mellor, G. L. (1987). A description of a three-dimensional coastal circulation model. In *Three Dimensional Coastal Ocean Models*, Coastal and Estuarine Sci., Ser. 4, ed. N. Heaps. Washington, DC: American Geophysical Union, pp. 1–16.

Bogue, N., Huang, R.-X. and Bryan, K. (1986). Verification experiments with an isopycnal coordinate model. *J. Phys. Oceanogr.* **16**, 985–90.

Böning, C. W., Döscher, R. and Budich, R. G. (1991). Seasonal transport variation in the western subtropical north Atlantic: Experiments with an eddy-resolving model. *J. Phys. Oceanogr.*, **21**, 1271–89.

Boulanger, J.-P., Delecluse, P., Maes, C. and Lévy, C. (1997). Long equatorial waves in a high-resolution OGCM simulation of the tropical Pacific ocean during the 1985–94 TOGA period. *Mon. Wea. Rev.*, **125**, 972–84.

Brink, K.H. (1991). Coastal-trapped waves and wind-driven currents over the continental shelf. *Ann. Rev. Fluid Mech.*, **23**, 389–412.

Bryan, F. O., Smith, R. D., Maltrud, M. E. and Hecht, M. W. (1998). Modeling the North Atlantic Circulation: from Eddy Permitting to Eddy Resolving. International WOCE Conference on Ocean Circulation and Climate, Halifax, 28 May 1998.

Bryan, F. (1987). Parameter sensitivity of primitive equation ocean circulation models. *J. Phys. Oceanogr.*, **17**, 970–85.

Bryan, K. (1963). A numerical investigation of a nonlinear model of a wind-driven ocean. *J. Atmos. Sci.*, **20**, 594–606.

Bryan, K. (1969). A numerical method for study of the circulation of the World Ocean. *J. Comput. Phys.*, **4**, 347–76.

Bryan, K. and Cox, M. D. (1967). A numerical investigation of the oceanic general circulation. *Tellus*, **19**, 54–90.

Busalacchi, A. J. and O'Brien, J. J. (1980). The seasonal variability in a model of the tropical Pacific. *J. Phys. Oceanogr.*, **10**, 1929–51.

Busalacchi, A. J. and O'Brien, J. J. (1981). Interannual variability of the equatorial Pacific. *J. Geophys. Res.*, **86**, 10901–7.

Cane, M. A. (1984). Modeling sea level during El Niño. *J. Phys. Oceanogr.*, **14**, 1864–74.

Cane, M. A. and Patton, R. J. (1984). A numerical model for low-frequency equatorial dynamics. *J. Phys. Oceanogr.*, **14**, 1853–63.

Cane, M. A. and Sarachik, E. S. (1977). Forced baroclinic ocean motions. II. The linear equatorial bounded case. *J. Mar. Res.*, **35**, 395–432.

Cane, M. A. and Sarachik, E. S. (1981). The response of a linear baroclinic equatorial ocean to periodic forcing. *J. Mar. Res.*, **39**, 651–93.

Chan, N. H., Kadane, J. B., Miller, R. N. and Palma, W. (1996). Predictions of tropical sea level anomaly by an improved Kalman filter. *J. Phys. Oceanogr.*, **26**, 1286–303.

Chapman, D. C. (2000). A numerical study of the adjustment of a narrow stratified current over a sloping bottom. *J. Phys. Oceanogr.*, **30**, 2927–40.

Chapman, D. C. and Lentz, S. J. (1997). Adjustment of stratified flow over a sloping bottom. *J. Phys. Oceanogr.*, **27**, 340–56.

Charney, J. G. (1955). The generation of oceanic currents by wind. *J. Marine Res.*, **14**, 477–98.

Charney, J. G., Fjortoft, R. and von Neumann, J. (1950). Numerical integration of the barotropic vorticity equation. *Tellus*, **2**, 237–54.

Chassignet, E. P., Arango, H., Dietrich, D., Ezer, T., Ghil, M., Haidvogel, D. B., Ma, C.-C., Mehra, A., Paiva, A. and Sirkes, Z. (2000). DAMÉE-NAB: the base experiments. *Dyn. Atmos. Oceans*, **32**, 155–83.

Chelton, D. B., Schlax, M. G., Lyman, J. M. and Johnson, G. C. (2003). Equatorially trapped Rossby waves in the presence of a meridionally sheared baroclinic flow in the Pacific Ocean. *Prog. Oceanogr.*, **56**, 323–80.

Chen, D., Liu, W. T., Zebiak, S. E., Cane, M. A., Kushnir, Y. and Witter, D. (1999). Sensitivity of the tropical Pacific Ocean simulation to the temporal and spatial resolution of wind forcing. *J. Geophys. Res.*, **104**, 11261–71.

Chorin, A. J. (1994). *Vorticity and Turbulence.* New York: Springer-Verlag.

Copson, E. T. (1962). *Theory of Functions of a Complex Variable.* Oxford: Oxford University Press.

Cox, M. D. (1984). *A Primitive Equation, 3-dimensional Model of the Ocean*, GFDL Ocean Group Tech. Rep. No. 1, GFDL/Princeton University.

Cox, M. D. (1987). Isopycnal diffusion in a z-coordinate ocean model. *Ocean Modelling*, Issue 74, 1–5.

Cox, M. D. (1989). An idealized model of the world ocean. Part I: The global scale water masses. *J. Phys. Oceanogr.*, **19**, 1730–52.

da Silva, A. M., Young, C. C. and Levitus, S. (1994). *Atlas of Surface Marine Data 1994, Vol. 3, Anomalies of Heat and Momentum Fluxes.* NOAA Atlas, NESDIS 8. Washington, DC: NOAA.

Davies, H. C. (1973). On the initial-boundary value problem of some geophysical fluid flows. *J. Comput. Phys.*, **13**, 398–422.

Déqué, M., Dreveton, C., Braun, A. and Cariolle, D. (1994). The ARPEGE/IFS atmosphere model: a contribution to the French community climate modeling. *Climate Dyn.*, **10**, 249–66.

de Szoeke, R. A. (1998). Equations of motion using thermodynamic coordinates. *J. Phys. Oceanogr.*, **30**, 2814–29.

Dietrich, D. E. (1997). Application of a modified A-grid ocean model having reduced numerical dispersion to the Gulf of Mexico circulation. *Dyn. Atmos. Oceans*, **27**, 201–317.

Dijkstra, H. A. and Ghil, M. (2005). Low-frequency variability of the large-scale ocean circulation: a dynamical systems approach. *Rev. Geophys.*, **43**, 122–59.

Dijkstra, H. A. and Molemaker, M. J. (1999). Imperfections of the North Atlantic wind-driven ocean circulation: continental geometry and wind stress shape. *J. Mar. Res.*, **57**, 1–28.

Döscher, R., Böning, C. W. and Herrmann, P. (1994). Response of circulation and heat transport in the north Atlantic to changes in thermohaline forcing in northern latitudes: a model study. *J. Phys. Oceanogr.*, **24**, 2306–20.

Dukowicz, J. K. and Smith, R. D. (1994). Implicit free-surface method for the Bryan–Cox–Semtner ocean model. *J. Geophys. Res.*, **99**, 7991–8014.

du Penhoat, Y. and Treguier, A. M. (1985). The seasonal linear response of the tropical Atlantic Ocean. *J. Phys. Oceanogr*, **15**, 316–29.

du Penhoat, Y., Cane, M. A. and Patton, R. J. (1983). Reflection of low-frequency equatorial waves on partial boundaries. In *Hydrodynamics of the Equatorial Ocean*, ed. J. C. J. Nihoul. Amsterdam: Elsevier, pp. 237–58.

Durran, D. (1999). *Numerical Methods for Wave Equations in Geophysical Fluid Dynamics.* New York: Springer-Verlag.

Engquist, B. and Majda, A. (1977). Absorbing boundary conditions for the numerical simulation of waves. *Math. Comp.*, **31**, 629–51.

Federiuk, J. and Allen, J. S. (1995). Upwelling circulation on the Oregon continental shelf. Part II: Simulations and comparisons with observations. *J. Phys. Oceanogr.*, **25**, 1867–89.

Fletcher, C. A. J. (1991). *Computational Techniques for Fluid Dynamics.* 2 vols. Berlin: Springer-Verlag.

Foreman, M. G. G., Crawford, W. R., Cherniawsky, J. Y., Henry, R. F. and Tarbotton, M. (2000). A high-resolution assimilating tidal model for the Northeast Pacific Ocean. *J. Geophys. Res.*, **105**, 28629–51.

Foreman, M. G. G., Henry, R. F., Walters, R. A. and Ballantyne, V. A. (1993). A finite element model for tides and resonance along the north coast of British Columbia. *J. Geophys. Res.*, **98**, 2509–32.

Gan, J. and Allen, J. S. (2002). A modeling study of shelf circulation off northern California in the region of the Coastal Ocean Dynamics Experiment: response to relaxation of upwelling winds. *J. Geophys. Res.*, **107**, 3123–53.

Gates, W. L. (1968). A numerical study of transient Rossby waves in a wind-driven homogeneous ocean. *J. Atmos. Sci.*, **25**, 3–22.

Gerdes, R. (1993). A primitive equation ocean circulation model using a general vertical coordinate transformation. 1. Description and testing of the model. *J. Geophys. Res.*, **98**, 14683–701.

Gordon, A. L. and Greengrove, C. (1986). Geostrophic circulation of the Brazil-Falkland confluence. *Deep-Sea Res.*, **33**, 573–85.

Griffies, S. M. (1998). The Gent–McWilliams skew-flux. *J. Phys. Oceanogr.*, **28**, 831–41.

Griffies, S. M., Gnandesikan, A., Pacanowski, R. C., Larichev, V., Dukowicz, J. K. and Smith, R. D. (1998). Isoneutral diffusion in a z-coordinate ocean model. *J. Phys. Oceanogr.*, **28**, 805–30.

Guckenheimer, J. and Holmes, P. (1983). *Nonlinear Oscillations, Dynamical Systems and Bifurcations of Vector Fields.* Applied Mathematical Sciences, 42. New York: Springer-Verlag.

Hackert, E. C., Busalacchi, A. J. and Murtugudde, R. (2001). A wind comparison study using an ocean general circulation model for the 1997–1998 El Niño. *J. Geophys. Res.*, **106**, 2345–62.

Haidvogel, D. B. and Beckmann, A. (1999). *Numerical Ocean Circulation Modeling.* London: Imperial College Press.

Haidvogel, D. B., Robinson, A. R. and Schulman, E. E. (1980). The accuracy, efficiency and stability of three numerical models with application to open-ocean problems. *J. Comput. Phys.*, **34**, 1–53.

Haidvogel, D. B., Wilkin, J. L. and Young, R. E. (1991). A semi-spectral primitive equation ocean circulation model using vertical sigma and orthogonal curvilinear coordinates. *J. Comput. Phys.*, **94**, 151–85.

Halliwell, G. R. and Allen, J. S. (1984). Large-scale sea level response to atmospheric forcing along the west coast of north America, summer 1973. *J. Phys. Oceanogr.*, **14**, 864–86.

Haltiner, G. J. and Williams, R. T. (1980). *Numerical Prediction and Dynamic Meteorology*, 2nd edn. New York: John Wiley and Sons.

Haney, R. L. (1991). On the pressure gradient force over steep topography in sigma coordinate ocean models. *J. Phys. Oceanogr.*, **21**, 610–19.

Hellerman, S. and Rosenstein, M. (1983). Normal monthly wind stress over the world ocean with error estimates. *J. Phys. Oceanogr.*, **13**, 1093–104.

Higdon, R. L. (1986). Initial-boundary value problems for linear hyperbolic systems. *SIAM Rev.*, **28**, 177–217.

Higdon, R. L. (1994). Radiation boundary conditions for dispersive waves. *SIAM J. Numer. Anal*, **31**, 64–100.

Hogg, A. M., Dewar, W. K., Killworth, P. D. and Blundell, J. R. (2003). A quasi-geostrophic coupled model (Q-GCM). *Mon. Wea. Rev.*, **131**, 2261–78.

Holton, J. R. (1992). *An Introduction to Dynamic Meteorology*, 3rd edn. San Diego: Academic Press.

Irons, B. and Shrive, N. (1983). *Finite Element Primer.* Chichester: Ellis Horwood Ltd.

Isaacson, E. and Keller, H. B. (1966). *Analysis of Numerical Methods*, New York: John Wiley and Sons.

Jespersen, D. C. (1974). Arakawa's method is a finite-element method. *J. Comput. Phys.*, **16**, 383–90.

Jiang, S., Jin, F.-f. and Ghil, M. (1995). Multiple equilibria, periodic and aperiodic solutions in a wind-driven, double-gyre, shallow-water model. *J. Phys. Oceanogr.*, **25**, 764–86.

Kawabe, M. (1994). Mechanisms of interannual variations of equatorial sea level associated with El Niño. *J. Phys. Oceanogr.*, **24**, 979–93.

Killworth, P. D. (1987). Topographic instabilities in level model OGCMs. *Ocean Modelling*, No. 75, 9–12.

Kreiss, H.-O. (1970). Initial boundary value problems for hyperbolic systems. *Comm. Pure Appl. Math.*, **23**, 277–98.

Lee, T. N., Johns, W. E., Zantopp, R. J. and Fillenbaum, E. R. (1996). Moored observations of western boundary current variability and thermocline circulation at 26.5° N in the subtropical north Atlantic. *J. Phys. Oceanogr.*, **26**, 962–83.

Le Provost, C., Genco, M. L., Lyard, F., Vincent, P. and Canceil, P. (1994). Spectroscopy of the world tides from a finite element hydrodynamic model. *J. Geophys. Res.*, **99**, 24777–97.

Leveque, R. J. (1992). *Numerical Methods for Conservation Laws*, 2nd edn. Basel: Birkhauser Verlag.

Levitus, S. (1982). *Climatological Atlas of the World Ocean.* NOAA Prof. Paper, 13, Washington, DC: NOAA.

Levitus, S. and Gelfeld, R. (1992). *NODC Inventory of Physical Oceanographic Profiles, Key to Oceanographic Records Documentation*, vol. 18, Natl. Oceanic Data Cent., Washington, DC.

Liu, W. T., Tang, W. and Polito, P. S. (1998). NASA scatterometer provides global ocean-surface wind fields with more structures than numerical weather prediction. *Geophys. Res. Lett*, **25**, 761–4.

Lynch, D. R. and Holbroke, M. J. (1997). Normal flow boundary conditions in 3D circulation models. *Int. J. Numer. Meth. Fluids*, **25**, 1185–205.

Lynch, D. R., Ip, J. T. C., Naimie, C. E. and Werner, F. E. (1996). Comprehensive coastal circulation model with application to the Gulf of Maine. *Continental Shelf Res.*, **16**, 875–906.

MacCready, P. and Rhines, P. B. (1993). Slippery bottom boundary layers on a slope, *J. Phys. Oceanogr.*, **23**, 5–22.

Maes, C., Madec, G. and Delecluse, P. (1997). Sensitivity of an equatorial Pacific OGCM to the lateral diffusion. *Mon. Wea. Rev.*, **125**, 958–71.

Maltrud, M., Smith, R. D., Semtner, A. J. and Malone, R. C. (1998). Global eddy-resolving ocean simulations driven by 1985–1995 atmospheric winds. *J. Geophys. Res.*, **103**, 30825–53.

Marshall, J., Adcroft, A., Hill, C., Perelman, L. and Heisey, C. (1997a). A finite-volume, incompressible Navier-Stokes model for studies of the ocean on parallel computers. *J. Geophys. Res.*, **102**, 5733–52.

Marshall, J., Hill, C., Perelman, L. and Adcroft, A. (1997b). Hydrostatic, quasi-hydrostatic and non-hydrostatic ocean modeling. *J. Geophys. Res.*, **102**, 5733–52.

Matano, R. P. (1993). On the separation of the Brazil current from the coast. *J. Phys. Oceanogr.*, **23**, 79–90.

McCalpin, J. D. (1994). A comparison of second-order and fourth-order pressure gradient algorithms in a σ-coordinate ocean model. *Int. J. Numer. Meth. Fluids*, **18**, 361–83.

McClean, J. L., Semtner, A. J. and Zlotnicki, V. (1997). Comparisons of mesoscale variability in the Semtner-Chervin $1/4°$ model, the Los Alamos Parallel Ocean Program $1/6°$ model, and TOPEX/POSEIDON data. *J. Geophys. Res.*, **102**, 25203–26.

McDougall, T. J., (1987). Neutral surfaces. *J. Phys. Oceanogr.*, **17**, 1950–64.

Mellor, G. L., Ezer, T. and Oey, L.-Y. (1994). The pressure gradient conundrum of sigma coordinate ocean models. *J. Atmos. Oceanic Technol.*, **11**, 1126–34.

Mellor, G. L. and Yamada, T. (1982). Development of turbulence closure model for geophysical fluid problems. *Rev. Geophys. Space Phys.*, **20**, 851–75.

Mesinger, F. and Arakawa, A. (1976) *Numerical Methods used in Atmospheric Models.* GARP Publication Series No. 14, WMO/ICSU Joint Organizing Committee.

Miller, R. N. (1986). Toward the application of the Kalman filter to regional open-ocean modeling. *J. Phys. Oceanogr.*, **16**, 72–86.

Miller, R. N. and Bennett, A. F. (1988). Numerical simulation of flows with locally characteristic boundaries. *Tellus*, **40A**, 303–23.

Miller, R. N. and Cane, M. A. (1989). A Kalman filter analysis of sea-level height in the tropical Pacific. *J. Phys. Oceanogr.*, **19**, 773–90.

Miller, R. N. and Cane, M. A. (1997). Tropical data assimilation: theoretical aspects. In *Modern Approaches to Data Assimilation in Ocean Modeling*, ed. P. Malanotte-Rizzoli. Elsevier Oceanography Series, 61. Amsterdam: Elsevier, pp. 207–33.

Miller, R. N., Robinson, A. R. and Haidvogel, D. B. (1983). A baroclinic quasigeostrophic open-ocean model. *J. Comput. Phys.*, **50**, 38–70.

Munk, W. H. (1950). On the wind-driven ocean circulation. *J. Meteorology*, **7**, 79–93.

Neelin, J. D., Battisti, D. S., Hirst, A. C., Jin, F.-F., Wakata, Y., Yamagata, T. and Zebiak, S. E. (1998). ENSO theory. *J. Geophys. Res.*, **103**, 14261–90.

Oberhuber, J. M. (1988). An atlas based on the COADS data set: the budgets of heat, buoyancy, and turbulent kinetic energy at the surface of the global ocean. Rep. 15, Max-Planck-Institut für Meteorologie, Bundesstrasse 55, 20146 Hamburg, Germany.

Oliger, J. and Sundstrom, A. (1978). Theoretical and practical aspects of some initial boundary value problems in fluid dynamics. *SIAM J. Appl. Math.*, **35**, 419–46.

Orlanski, I. (1976). A simple boundary condition for unbounded hyperbolic flows. *J. Comput. Phys.*, **21**, 251–69.

Pacanowski, R. C. and Philander, S. G. H. (1981). Parameterization of vertical mixing in numerical models of the tropical ocean. *J. Phys. Oceanogr.*, **11**, 1443–51.

Patera, A. T. (1984). A spectral element method for fluid dynamics: laminar flow in a channel expansion. *J. Comput. Phys.*, **54**, 468–88.

Pedlosky, J. (1979). *Geophysical Fluid Dynamics.* New York: Springer-Verlag.

Pedlosky, J. (1987). An inertial theory of the equatorial undercurrent. *J. Phys. Oceanogr.*, **17**, 1978–85.

Perez, R. C. (2005). Numerical and assimilative studies of the equatorial Pacific cold tongue. Unpublished doctoral dissertation, Oregon State University.

Perez, R. C., Chelton, D. B. and Miller, R. N. (2005). The effects of wind forcing and background mean currents on the latitudinal structure of equatorial Rossby waves. *J. Phys. Oceanogr.*, **35**, 666–82.

Peterson, R. G., Johnson, C. S., Krauss, H. and Davis, R. E. (1996). Lagrangian measurements in the Malvinas Current. In *The South Atlantic: Present and Past Circulation*, eds. G. Wefer, W. Berger, G. Siedler and D. J. Webb. Berlin; New York: Springer-Verlag, pp. 239–247.

Philander, S. G. H. (1991). *El Niño, La Niña and the Southern Oscillation.* San Diego: Academic Press.

Philander, S. G. H. and Pacanowski, R. C. (1984). A model of the seasonal cycle in the tropical Atlantic ocean. *Geophys. Res. Lett.*, **11**, 802–4.

Philander, S. G. H. and Pacanowski, R. C. (1986a). A model of the seasonal cycle in the tropical Atlantic ocean. *J. Geophys. Res.*, **91**, 14192–206.

Philander, S. G. H. and Pacanowski, R. C. (1986b). The mass and heat budget in a model of the tropical Atlantic ocean. *J. Geophys. Res.*, **91**, 14212–20.

Philander, S. G. H., Hurlin, W. J. and Pacanowski, R. C. (1986). Properties of long equatorial waves in models of the seasonal cycle in the tropical Atlantic and Pacific oceans. *J. Geophys. Res.*, **91**, 14207–11.

Ralston, A. (1965). *A First Course in Numerical Analysis.* New York: McGraw-Hill.

Ray, R. D. (1993). Global ocean tide models on the eve of TOPEX/POSEIDON. *IEEE Trans. Geoscience and Remote Sensing*, **31**, 355–64.

Rebert, J. P., Donguy, J. R., Eldin, G. and Wyrtki, K. (1985). Relations between sea level, thermocline depth, heat content and dynamic height in the tropical Pacific ocean. *J. Geophys. Res.*, **90**, 11719–25.

Redi, M. (1982). Oceanic isopycnal mixing by coordinate rotation. *J. Phys. Oceanogr.*, **12**, 1154–8.

Reynolds, R. W. and Smith, T. M. (1994). Improved global sea surface temperature analysis using optimum interpolation. *J. Clim.*, **7**, 929–48.

Richardson, L. F. (1965). *Weather Prediction by Numerical Process.* New York: Dover.

Richardson, P. L. and McKee, T. K. (1984). Average seasonal variation of the Atlantic north equatorial countercurrent from ship drift data. *J. Phys. Oceanogr.*, **14**, 1226–38.

Richtmyer, R. D. (1963). A survey of difference methods for non-steady fluid dynamics. *NCAR Tech. Notes 63-2*, National Center for Atmospheric Research, Boulder, CO.

Richtmyer, R. D. and Morton, K. W. (1967). *Difference Methods for Initial Value Problems.* New York: Interscience.

Robinson, A. R. and Walstad, L. J. (1987) The Harvard open-ocean model: calibration and application to dynamical process, forecasting and data assimilation studies. *Appl. Numer. Math.*, **3**, 89–131.

Sadourny, R. (1975). The dynamics of finite-difference models of the shallow-water equations. *J. Atmos. Sci.*, **32**, 680–9.

Sarmiento, J. L. and Bryan, K. (1982). An ocean transport model for the North Atlantic. *J. Geophys. Res.*, **87**, 394–408.

Schlax, M. G., Chelton, D. B. and Freilich, M. H. (2001). Sampling errors in wind fields constructed from single and tandem scatterometer datasets. *J. Atmos. Oceanic Technol.*, **18**, 1014–36.

Schmitz, W. J. and Thompson, J. D. (1993). On the effects of horizontal resolution in a limited-area model of the Gulf Stream system. *J. Phys. Oceanogr.*, **23**, 1001–7.

Semtner, A. J. Jr. and Chervin, R. M. (1988). A simulation of the global ocean circulation with resolved eddies. *J. Geophys. Res.*, **93**, 15502–22.

Semtner, A. J. Jr., and Chervin, R. M. (1992). Ocean general circulation from a global eddy-resolving model. *J. Geophys. Res.*, **97**, 5493–550.

Semtner, A. J. Jr. and Holland, W. R. (1978). Intercomparison of quasi-geostrophic simulations of the western north Atlantic circulation with primitive equation results. *J. Phys. Oceanogr.*, **8**, 735–54.

Shapiro, R. (1970). Smoothing, filtering, and boundary effects. *Rev. Geophys. Space. Phys.*, **8**, 359–87.

Smagorinsky, J. (1963) General circulation experiments with the primitive equations I. The basic experiment. *Mon. Wea. Rev.*, **91**, 99–164.

Smedstad, O. M. and O'Brien, J. J. (1991). Variational data assimilation and parameter estimation in an equatorial Pacific ocean model. *Prog. Oceanogr.*, **26**, 179–241.

Smith, R. D., Maltrud, M. E., Bryan, F. O. and Hecht, M. W. (2000). Numerical simulation of the north Atlantic at $1/10°$. *J. Phys. Oceanogr.*, **30**, 1532–60.

Smith, R. L., (1981). A comparison of the structure and variability of the flow field in three coastal upwelling regions: Oregon, Northwest Africa and Peru Coastal Upwelling. *Coastal and Estuarine Sciences*, **1**, 107–118.

Sod, G. A. (1985). *Numerical Methods in Fluid Dynamics.* New York: Cambridge University Press.

Song, Y. and Haidvogel, D. B. (1994). A semi-implicit ocean circulation model using a generalized topography-following coordinate system. *J. Comput. Phys.*, **94**, 151–85.

Spall, M. A. (1988). Regional ocean modeling: primitive equation and quasi-geostrophic studies. Doctoral dissertation, published as Harvard University Reports in Meteorology and Oceanography no. 28, Harvard University, Cambridge, MA.

Spall, M. A. and Robinson, A. R. (1990). Regional primitive equation studies of the Gulf Stream meander and ring formation region. *J. Phys. Oceanogr.*, **20**, 985–1016.

Speich, S., Dijkstra, H. A. and Ghil, M. (1995). Successive bifurcations of a shallow-water model with applications to the wind-driven circulation. *Nonlin. Proc. Geophys.*, **2**, 241–68.

Stammer, D., Tokmakian, R., Semtner, A. and Wunsch, C. (1996). How well does a 1/4° global circulation model simulate large-scale oceanic observations? *J. Geophys. Res.*, **101**, 25779–811.

Stommel, H. M. (1966). *The Gulf Stream*, 2nd edn. Berkeley; Los Angeles: University of California Press.

Strang, G. and Fix, G. J. (1973) *An Analysis of the Finite Element Method.* New Jersey: Prentice-Hall, Inc.

Stricherz, J. N., O'Brien, J. J. and Legler, D. M. (1992). *Atlas of Florida State University Tropical Pacific Winds for TOGA*, Mesoscale Air-Sea Interaction Group technical report, Florida State University, Tallahassee, FL.

Sun, S., Bleck, R., Rooth, C., Dukowicz, J., Chassignet, E. and Killworth, P. (1999). Inclusion of thermobaricity in isopycnic-coordinate ocean models. *J. Phys. Oceanogr.*, **29**, 2719–29.

Sundstrom, A. (1969). Stability theorems for the barotropic vorticity equation. *Mon. Wea. Rev.*, **97**, 340–5.

Sura, P., Fraedrich, K. and Lunkeit, F. (2001). Regime transitions in a stochastically forced double-gyre model. *J. Phys. Oceanogr.*, **31**, 411–26.

Talley, L. D., Reid, J. L. and Robbins, P. E. (2003). Data-based meridional overturning streamfunctions for the global ocean. *J. Clim.*, **16**, 3213–26.

Toggweiler, J. R. K., Dixon, K. and Bryan, K. (1989). Simulations of radiocarbon in a coarse-resolution world ocean model 1. Steady state prebomb conditions. *J. Geophys. Res.*, **94**, 8217–42.

Wacongne, S. (1989). Dynamical regimes of a fully nonlinear stratified model of the Atlantic equatorial undercurrent. *J. Geophys. Res.*, **94**, 4810–15.

Wacongne, S., (1990). On the difference in strength between Atlantic and Pacific undercurrents. *J. Phys. Oceanogr.*, **20**, 792–799.

Wajsowicz, R. C. (1986). Free planetary waves in finite-difference numerical models. *J. Phys. Oceanogr.*, **16**, 773–89.

Wajsowicz, R. C. (1993). A consistent formulation of the anisotropic stress tensor for use in models of the large-scale ocean circulation. *J. Comput. Phys.*, **105**, 333–8.

Wakata, Y. and Sarachik, E. S. (1991). On the role of equatorial ocean modes in the ENSO cycle. *J. Phys. Oceanogr.*, **21**, 434–43.

Weaver, A. J. and Sarachik, E. S. (1990). On the importance of vertical resolution in certain ocean general circulation models. *J. Phys. Oceanogr.*, **20**, 600–9.

Weisberg, R. H., Horigan, A. and Colin, C. (1979). Equatorially trapped Rossby–Gravity wave propagation in the Gulf of Guinea. *J. Mar. Res.*, **37**, 67–86.

Wentz, F. J., Gentemann, C., Smith, D. and Chelton, D. (2000). Satellite measurements of sea surface temperature through clouds. *Science*, **288**, 847–50.

Whitham, G. B. (1974). *Linear and Nonlinear Waves*, New York: Wiley-Interscience.

Willebrand, J., Barnier, B., Böning, C., Dieterich, C., Killworth, P. D., Le Provost, C., Jia, Y., Molines, J.-M. and New, A. L. (2001). Circulation characteristics in three eddy-permitting models of the north Atlantic. *Prog. Oceanogr.*, **48**, 123–61.

Zebiak, S. E. and Cane, M. A., (1987). A model El Niño-Southern Oscillation. *Mon. Wea. Rev.*, **115**, 2262–78.

Zienkiewicz, O. C., (1977). *The Finite Element Method*, 3d expanded and revised edn. London; New York: McGraw-Hill (UK) Limited.

Index

Printed in the United States
By Bookmasters